The Southern Way

The regular volume for the Southern devotee

Kevin Robertson

Issue 53

www.crecy.co.uk

© 2021 Crécy Publishing Ltd
and the various contributors

ISBN 9781800350212

First published in 2021 by Noodle Books
an imprint of Crécy Publishing Ltd

New contact details
All editorial submissions to:
The Southern Way (Kevin Robertson)
'Silmaril'
Upper Lambourn
Hungerford
Berkshire RG17 8QR
Tel: 01488 674143
editorial@thesouthernway.co.uk

All rights reserved. No part of this book may be reproduced or transmitted in any form or by any means electronic or mechanical, including photocopying, recording or by any information storage without permission from the Publisher in writing. All enquiries should be directed to the Publisher.

A CIP record for this book is available from the British Library

Publisher's note: Every effort has been made to identify and correctly attribute photographic credits. Any error that may have occurred is entirely unintentional.

Printed in Malta by Melita Press

Noodle Books is an imprint of
Crécy Publishing Limited
1a Ringway Trading Estate
Shadowmoss Road
Manchester M22 5LH

www.crecy.co.uk

Front cover:
Amidst the general detritus of the running shed, S15 No. 30843 stands outside the front of Salisbury shed on an unreported date. The discs are obviously applicable to a West of England duty, although the 'SPL 5' on the lower disc is a bit of a mystery. Obviously prepared for its next turn, the coal on the tender looks more a mixture of dust and smalls, with that on No. 34099 *Lynmouth* alongside perhaps slightly better. The presence of the fire devils might indicate winter time but if this were the case escaping steam would probably also be more apparent. Inside the shed at least one other Bulleid type is present, although in reality there were probably several other engines within. No. 30843 is in typical grimy condition, with 'Meldon dust' visible on the cylinders, while the streaks in the side of the firebox are clear evidence of priming. Hopefully she will have received a recent boiler wash out at least.

Rear cover:
Former Bulleid open-third No. E1505 but seen here with Eastern Region identification (SR number retained but 'E' prefix added) and attached to another former Southern Region Bulleid, this one No. E1472 behind. Twenty Bulleid vehicles were exchanged between the Southern and Eastern/Scottish regions in exchange for Mk1 stock, the latter intended to go towards the REP/TC conversions. Slightly surprising is that these twenty vehicles remained the responsibility of the Southern Region for major maintenance; not that many (if any) ever returned there. If such attention was needed, the coach would have been withdrawn. For No. 1505, transfer was in September 1965, with withdrawal in August 1968, while for coach 1472 the dates were September 1965 and April 1968 respectively, giving the window of opportunity for the picture.
Notes courtesy of Mike King

Title page:
'Over the hump at Feltham'. Caught in the moment of 'push and let go', G16 No. 30492 is engaged in the task that it was designed for, shunting at the vast Feltham complex. Likely this was the last cut of wagons from a train and the crew may well have the chance to relax briefly before their next call to push (or pull). (Not, of course, the other type of 'push-pull' [or even 'pull-push'] that we are familiar with in passenger workings.) The date has to be c.1948 with the mixed identification present. In 2020 after many years of neglect, the site of the former marshalling yard may at last find a new use as a depot for electric sets operating Waterloo to Reading services. See also www.ianvisits.co.uk/blog/2020/05/13/south-western-railways-new-rail-depot-at-feltham/amp

Contents

Issue No 54 of THE SOUTHERN WAY
ISBN 9781800350236
available in April 2021 at £14.95

To receive your copy the moment it is released, order in advance from your usual supplier, or it can be sent post-free (UK) direct from the publisher:

Crécy Publishing Ltd (Noodle Books)

1a Ringway Trading Estate, Shadowmoss Road, Manchester M22 5LH

Tel 0161 499 0024

www.crecy.co.uk

enquiries@crecy.co.uk

Introduction	4
The S. C. Townroe Archive – in Colour Part 2	5
Windsor Line Steam	13
Wandsworth Town to Putney	
The Railways of the Wandle Valley Part 1	21
(Some) USA Tanks Before the Southern	28
War Comes to Eastleigh	31
Down to Earth Part 3A Ex-LBSCR Stock	43
One Evening in May	52
A Gosport Accident and the Eastleigh Footbridge	62
The Esher Slotted Signal	64
...and Speaking of Signals...Three Curios...	67
Remember *Remembrance* April 1922 to June 1935	70
The Southern on Social Media Pigeons and People	73
Colour Interlude	77
Rebuilt The Letters and Comments pages	82
'The Southdown Venturer Railtour' and the End of Steam The Last Rites of Steam and Lines in Sussex Part 2	88
Book Review *Parsons & Prawns* and *Southern Style*	94

Introduction

Regular readers will be aware that for reasons of practicality I have got into the habit of preparing an issue of SW several months ahead. I do not seek plaudits for seemingly trying to be efficient, more like a realisation that one never quite knows what is around the corner. Each issue also continues to be a team effort – I could not produce anything without the help of all the contributors – although the final compilation is down to me and the team here (meaning she who provides the coffee and if I am lucky home-made biscuits), plus the dogs who will come and lie at my feet.

Being in effect then a one-person band does have its advantages but similarly its disadvantages. There is no committee as such to decide on content and instead this is based upon three precepts; what there is available, what may have been sat in the pile and really does need an airing, and finally what I (and I do seek opinions from like-minded outside friends on this one) think is perhaps slightly different and consequently unusual. I really do try to use material in the order it is received, although do please bear in mind that it can sometimes not be easy if similar pieces might otherwise appear in consecutive issues. Ongoing series, of course, are a separate matter.

For this the January 2021 copy – where does the time go – we are including some items that had been intended for SW52 but in which we simply ran out of space. For this reason too there is on this occasion there is no instalment from John Click but be assured it (he) will be back. One of the reasons for this particular omission is that back in the summer of 2020 I was invited to rewrite my 2007 *Leader* book. Click's recollections have added greatly to our knowledge of the project (and certainly not all have or will appear in SW) and if added to the recollections of Harry Attwell and others plus even more images I hope the result will be as comprehensive an account as it is possible to produce. Indeed, as I write this I have just closed the file at the end of what will be Chapter 4 and already we are up to 50,000 words. Publication is scheduled for mid-2021.

To return to SW, we are also delighted to welcome George Reeve to the author pool and while, as mentioned earlier, there may be no continuation of John Click for this issue there certainly is a continuation of Mike King's popular series on grounded coach bodies. Of course, I do get to see these things before they are published and I have to admit I have found the whole series absolutely fascinating, Both Mike King and George Reeve are officers within the South Western Circle, a superb organisation if you care for the history of that company. Similarly, the Brighton Circle and the Southern Railways Group (plus others) have worked tirelessly to record and preserve information that might otherwise be lost. If you are not a member of at least one, you should be. I am personally humbled every time I read one of their articles compiled by what can only be said to be extremely learned individuals.

The year 2020 will, of course, be recalled more than anything else as the year of the pandemic. Not for one moment am I belittling the terrible toll it has taken on individuals, families and friends, but I similarly suspect that like me, many of you have also had your outside activities curtailed. Visits to heritage lines, museums and archives have all suffered, while the Health & Safety police seem ever more to control our lives – and not just relative to Covid. I think we all realise the steam (and electric) railway we seek to portray in *SW* could never have existed in today's environment, neither do I seek to defend the past for, let us be fair, there were some truly appalling work practices that went on especially a century or more ago. However, as I have also said, if we are attempting to operate a heritage railway recreating the past but in the twenty-first century, then some things do have to change. I say this as I heard recently how on one heritage line in the summer of 2020 a fireman had to be relieved from his footplate duty as he was literally overcome by heat exhaustion and on the point of collapse. All credit to the railway concerned for identifying the issue and dealing with it promptly. The individual concerned did not want to report sick as he felt he would be letting his mate, the public and the railway down. Far better for a service be delayed or cancelled than a worst-case scenario occur.

No Health & Safety issues though in this edition and instead I hope you can sit down with your favourite cup or tipple and enjoy what follows. As I have said before, we – and this includes the Crecy team – enjoy what we do, and what makes it even more rewarding is that you seem to like it too.

Kevin Robertson

The S. C. Townroe Archive – in Colour
Part 2 (Part 1 appeared in SW52)

We have been absolutely delighted with the response to the first instalment in the colour series of SC Townroe material and it is with pleasure that we present this second selection. As before, SCT was, shall we say, 'economical' at times with his caption information, sometimes a few words written on the slide mount, at other times nothing. Only a few dates are also available. But we don't think that really matters; the actual imagery and content is such that few had the same opportunity.

As before, we would also like to thank the Townroe family for allowing access to the material.

Copies of the SCT scans – but only as they appear – are available for private or commercial use. Please enquire at editorial@thesouthernway.co.uk. Watch this space also for details of something else very special planned for 2021.

Motive power man – no details unfortunately – holding the remains of the tender water strainer from a Merchant Navy. I suspect many readers (me included) had never even seen such a device, although I certainly did know they existed. We know it was recorded in April or May 1959, while the original cause had been a broken spring on the tender. No location or loco number is given.

5

Associated with the previous view, here are parts of the actual spring. The newspaper is *The Sunday Times* for 19 April 1959 and the (Technical) 'Situations Vacant'.

Below and opposite: An incident recorded on the old Ringwood line ('Castleman's Corkscrew') near Ringwood during demolition in 1965. Wagons had been placed at intervals into which concrete sleepers were being loaded ready to be collected later. Unfortunately someone had forgotten to apply the handbrake hard enough on at least one, or perhaps the extra weight was too much for the amount of brake applied. Whatever, this was the result at Crow Crossing. With the railway no longer used, it is very unlikely the gates were replaced; it was just fortunate there was no road traffic at the time.

The S. C. Townroe Archive – in Colour

Three images now on a footplate trip between Verwood and Downton. These were taken from a BR Class 4 76xxx series engine and show the view ahead firstly at Verwood, then leaving Fordingbridge, and finally approaching the tunnel at Downton. In the latter image there is also evidence of what we may take is controlled burning of some sort, this based on the assessment there is a man standing nearby.

This page and overleaf: **We now have four views of 'Reconstruction at Branksome'; in 1966. They were taken sometime in 1966, so also soon after Bournemouth West had closed to passenger traffic and the site was being prepared for the new depot for electric stock. Seen are the old carriage sheds, subsequently re-clad and reused as the modern-day depot. The rusty rails form part of the west side of the triangle, whose viaduct remains today but the rails are absent.**

10

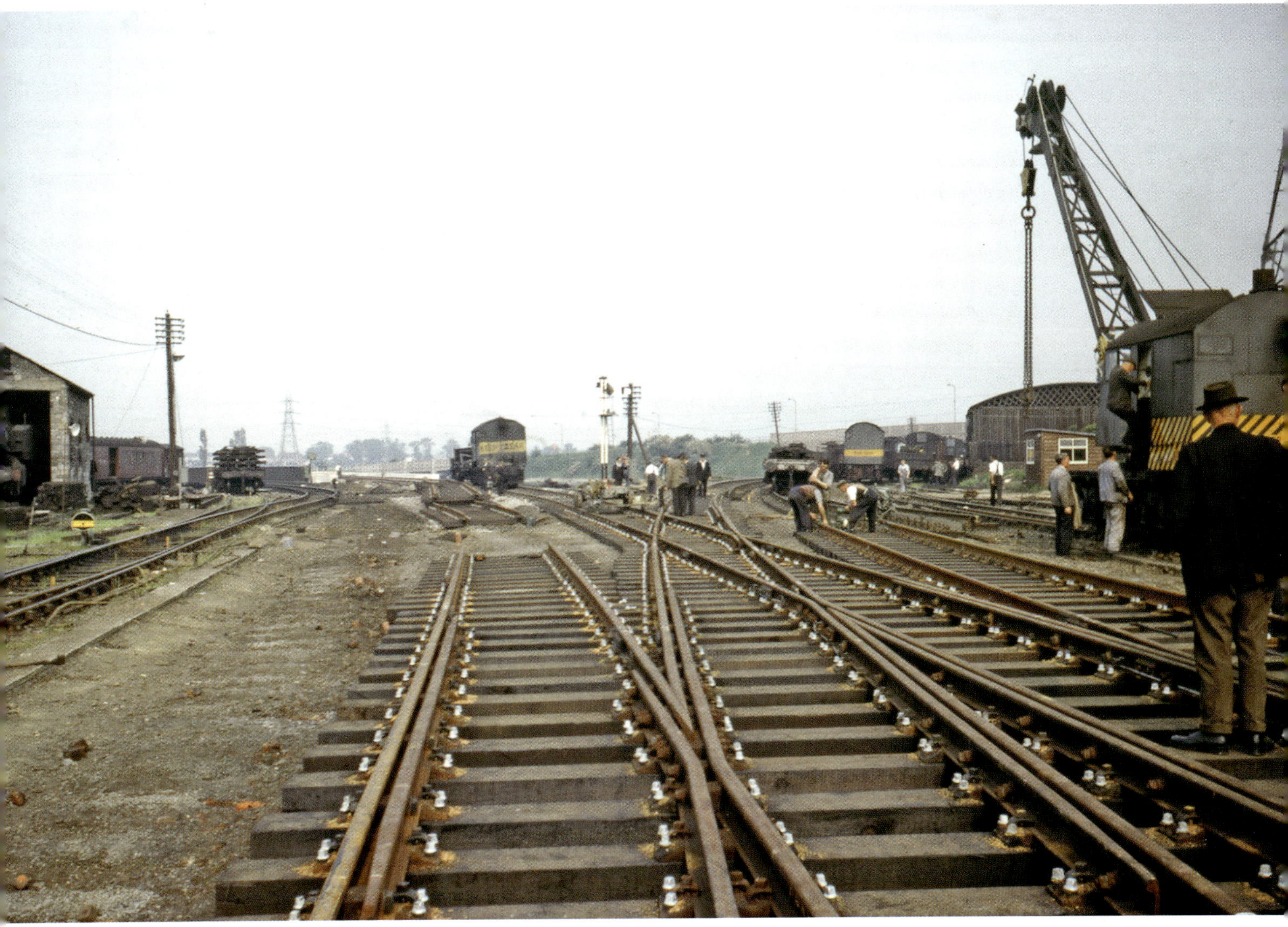

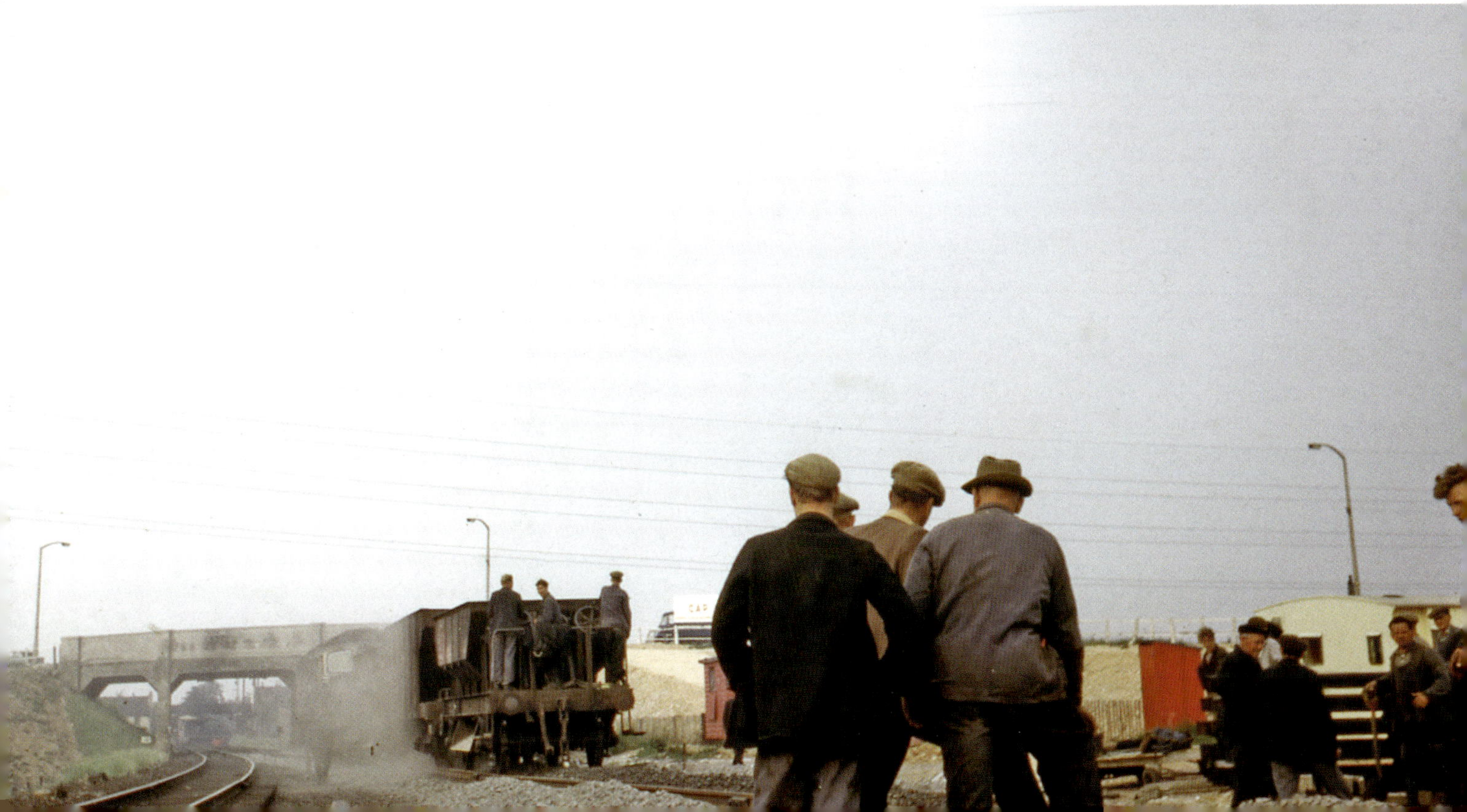

Above and previous page: **Interesting goings on at Redbridge on 31 March 1964, but first a brief description of the locality may not go amiss.** Redbridge is the junction of the Andover to Southampton line with the main line between Southampton and Bournemouth, the section between Andover and Romsey since closed, although that between Southampton and Romsey (and thence west to Salisbury) is still extremely busy. Redbridge was also the location of a sleeper works and, just beyond the junction and the works, a viaduct that carried the railway west across the estuary of the River Test. In the first image the new junction has been laid with workmen involved in connecting it to the existing rails. Notice, in this and the third view, the resident Redbridge depot USA tank skulking in its little shed. In the second image, ballasting and its associated dust is taking place behind a 'Crompton' diesel. The third view though is perhaps the most interesting as we have a Bulleid Pacific (we think No. 35025) cautiously making its way east on what is very temporary connection between the main line and the depot sidings in order to continue its journey to Southampton and Waterloo. Behind the engine the string of green coaches may be noted. Fine, so the train will only be proceeding at very low speed and there may consequently be a delay to the service but it is also another example of how the railway put passengers first, unlike today, the alternative of road transport being a last and not a first resort.

Next time: Breakdown duty at Reading ('Southern' of course), rebuilding No. 35014, clean engines, preserved engines, engines at the end of their life, and a diesel in trouble.

Windsor Line Steam
Wandsworth Town to Putney
Peter Tatlow

Heading east through Putney on the up local Windsor line, Ivatt 4MT No. 43019, originally built at Horwich with double chimney to Lot 188 in 1948 and withdrawn twenty years later, at the time allocated to Cricklewood, makes for Battersea Yard with a goods train from Brent via Barnes on 17 June 1959.

The LSWR's main line from Nine Elms, and later Waterloo, was from 1845 paralleled down to Clapham Jct, from where a line headed first for Richmond (1845) and later Windsor (1849), thereby giving rise to the term 'The Windsor Lines', and consequently never too far from the River Thames. In turn the line was extended from Staines to Reading (1856), while in the meantime a branch line from Barnes to beyond Twickenham gave access north of the Thames known as 'The Hounslow Loop' (1850). At Kew Bridge this was to make important connections to the north, such as Willesden on the LNWR and Brent (Cricklewood) on the Midland Railway, via the North & South Western Jct Rly (1853), and south to Nine Elms, Battersea Yard (LBSC) and Hither Green (SECR). This was to afford valuable routes for the transit of freight between the northern and southern companies, enhanced between the Western and Central and Eastern Divisions of the Southern Railway by the opening of Feltham hump marshalling yard following the First World War, all of which used the stretch of the Windsor lines through Putney.

The network in the neighbourhood was, however, as yet incomplete. In 1889 the Wimbledon and West Metropolitan line was opened, connecting Wimbledon with an end-on junction with the District line at Putney Bridge just north of the River Thames, crossing the Windsor Line at high level on a brick viaduct. In addition, a steeply graded connection was made between Point Pleasant Jct between Wandsworth Town and Putney on the now four-track Windsor line and East Putney to provide a diversionary route for the LSWR main line between Clapham and Wimbledon.

The widened lines from Clapham Jct to Barnes were configured from the north to the south: up local, up through, down through and down local, with the platforms arranged with a central island and two-side platforms. By the 1950s a reduction

13

Possibly a light-engine movement; No. 30524, H15 Class, built at Eastleigh in 1924, now with a 5,000-gallon bogie tender, was among the last of the class to be withdrawn in 1961. It is coupled to a Q1 in reverse, bearing discs indicating a train from Southampton Docks to Nine Elms via main line, perhaps Nine Elms motive power depot.

in staff could be achieved by having just one man on the central island platform, which meant swapping the roles of the through and local lines and running the faster trains on the locals, so that the stopping trains could call at the through platforms.

However, between Point Pleasant Jct and Putney there was a considerable curve on the down local adjacent to a mass concrete wall retaining the cutting slope supporting the East Putney line. By trimming back the slope, reducing the height of the retaining wall to below 3ft above rail level and thereby taking advantage of the reduction in lateral clearance from 5ft 3¾in (for new work) to 2ft 4¾in, i.e. platform height, to enable the curve to be slewed outwards, it was realised that the alignment of the curve could be improved. On a couple of days in July 1959, while undertaking a survey of the wall and cutting to explore this possibility, opportunities arose to take photographs of passing traffic while standing back as a train passed. In this case, it was especially interesting due to the large number of transfer freights on the way between the London Midland and Southern Regions.

Still on the up local, but looking west, Putney station is just visible beyond the second overbridge, the nearer one carrying the District line to Wimbledon. No. 48306 Stanier 8F built at Crewe in 1943 to Lot 159 and withdrawn 1966 has a long mineral train in tow from Brent to Hither Green Sidings via Factory Jct. Elsewhere, the lamp positions would suggest express train headlights, but the Southern Region used these to indicate the train's route and the Appendix to the Working Timetable contained special provision for foreign engines venturing into its territory in an attempt to clarify the intended route.

Later on the same day, Ivatt 4MT No. 43019 returns on the down through from Battersea Yard with a train of vans for Brent via Barnes passing under Woodlands Way overbridge with the flyover up connecting curve from East Putney beyond at a higher level. Note the engineer's level and tripod in the wideway.

This time a Fowler 4F No. 44297 from Cricklewood shed has another load of half a dozen trains during the day from Brent for Battersea Yard made up of a mix of open wagons and covered vans. No. 44297 was built at Derby in 1927 to Lot 29 and withdrawn in 1963.

The headcode would have us believe that No. 34060 *25 Squadron* heads a train from Plymouth to Waterloo or Nine Elms, but it is more likely empty carriage stock returning to Clapham Jct carriage sidings. This Battle of Britain Class No. 34060 was outshopped from Brighton in 1947 with a 4,500-gallon tender. It was rebuilt in November 1960, after which it survived until the end of steam on the Southern Region in July 1967.

Windsor Line Steam

In contrast to the Fowler 4F 0-6-0, Bulleid Q1 Class No. 33009 built at Brighton in 1942 and withdrawn in 1965, passes with a goods train on the down through line for Feltham, not going by Mortlake, but instead taking the Hounslow Loop.

No. 34023 *Blackmore Vale* passes beneath a District line train on its way to Wimbledon with what may well be another empty carriage working to Clapham Jct. No. 34023 was built in February 1946 also at Brighton. It survived to the end of steam, after which it was taken into preservation. On the left the chainman holds up the fully extended levelling staff.

No. 33015, another Q1 of 1942, speedily heads up the local line towards Clapham Jct with a transfer freight train fitted in between the tightly timed EMU service.

Once pride of the line, No. 30449 *Sir Torre,* an N15 'King Arthur' Class built at Eastleigh in 1925 with 5,000-gallon bogie tender and withdrawn at the end of 1959, hauls a train of mixed wagons and vans from Reading South for Nine Elms via Hounslow.

Windsor Line Steam

Right: With the surveying job incomplete, a return was made on 22 June 1959 when No. 30518, one of five Urie 4-6-2T H16 Class with 2,000-gallon water capacity built in 1922, was the first to pass with a train of containers for Feltham. The ventilated insulated containers were intended for the conveyance of fresh meat. Most are loaded on Conflats, although the fourth vehicle appears to be a brake-fitted three-plank drop-side wagon.

Below: **No. 30785** *Sir Mador de la Porte* was another N15, this time supplied by the North British Locomotive Co in 1925, with revised profile to the cab roof to suit the Eastern Section. When photographed it had just months to go before withdrawal. It is seen here with a train for Nine Elms from Reading South via Hounslow again passing beneath the viaduct carrying the District line.

No. 30699 Dugald Drummond 700 Class, known as 'Black Motors', was built by Dübs of Glasgow in 1897 and sometimes likened to the Jumbo Class he designed for the Caledonian Railway, especially prior to the installation of superheated boilers, but more akin to his brother Peter's 'Barney' Class for the Highland Railway. No. 30699 has a mineral train in tow; this time from Reading for Nine Elms having travelled via Hounslow.

Finally Hughes 'Crab' Class No. 42902 built as No. 13202 at Crewe in September 1930 to Lot 69 and withdrawn in February 1964 with a train from Brent to Hither Green via Barnes and Factory Jct.

References

Dendy Marshall, C F, *A History of the Southern Railway*, The Southern Railway, 1936.

Locomotives Western Section (Diagrams), British Railways, Southern Region, 1957–1962.

Miles, K, *Putney Freight June 1959*, British Railway Journal, Vol. 18, No. 4, Jan 2009, pp.170–173.

Rowledge, P, *Engines of the LMS built 1923–51*, Oxford Publishing Co, 1975.

Sectional Appendix to the working timetable and books of Rules and regulations, Western Section, British Railways, Southern Region, 1 October 1960, pp.120–124.

Tatlow, P, *Maintaining Railway Curves*, Backtrack, Vol 26, 2012, pp.123–125.

The Railways of the Wandle Valley
Part 1

Alan Postlethwaite

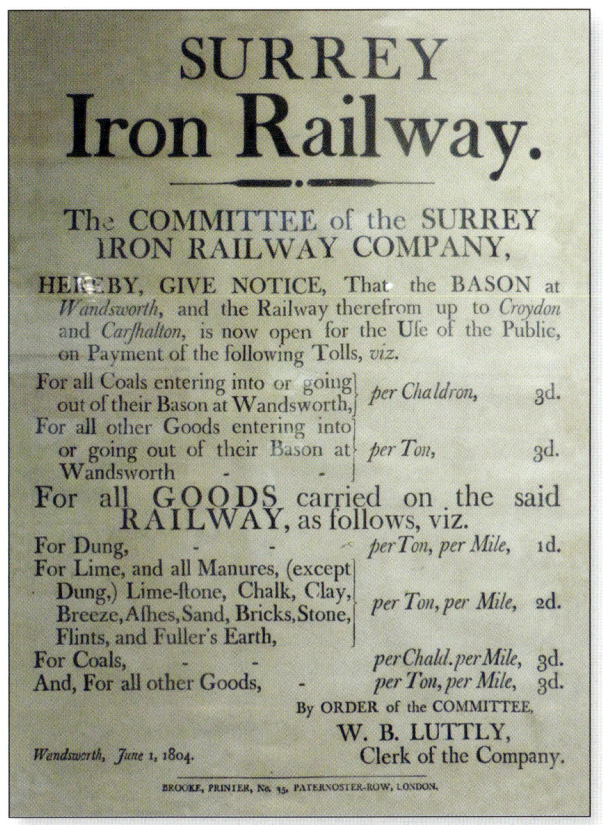

In June 2019, I walked the Wandle Trail, a walking and cycling path through linear parks from Wandsworth to the river's principal sources at Carshalton and Waddon. I photographed all the railway stations along the way and this article is the outcome, a cross section of suburbia. A full list of illustration credits is given at the end.

The Surrey Iron Railway

Over sixty water mills once exploited the River Wandle – a huge concentration of early industry. Canalisation was considered in 1799 but would have caused too much disruption to the mills. Instead, the Surrey Iron Railway was built, the world's first railway company and its first public railway. It was a freight-only toll line with the users providing their own wagons and their own horses, mules or donkeys. A list of tolls is shown in Figure 1. It was a plate railway with plain wheels and flanged rails to a gauge of 4ft 2in.

Fig. 2 shows some preserved samples of rail, stone sleepers, a wheel and a plaque.

A single horse might pull up to 55 tons of goods down the gentle grade from Croydon, then pull the empty wagons back up, perhaps with some coal. Wandsworth had many barge terraces for transloading goods. Fig. 3 shows one of them at low tide where the Wandle enters the Thames.

The Surrey Iron Railway followed the Wandle closely to Mitcham and then crossed Mitcham Common to Croydon with a short branch to Hackbridge. Just over 9 miles long, it opened in 1805. There was an end-on connection with the Croydon Merstham & Godstone Railway, which carried minerals from the Merstham area and the Greystone quarry. Competition arrived in 1809 when the Croydon Canal opened, connecting with the Thames at Rotherhithe. The canal closed in 1836, having been taken over by the London & Croydon Railway, which opened in 1839. The Surrey Iron Railway closed in 1846.

Fig. 1 *Alan Postlethwaite/Courtesy Bluebell Railway Museum*

Fig. 2 *Alan Postlethwaite/Wandle Industrial Museum, courtesy of the London Borough of Mitcham*

Fig. 3 *Alan Postlethwaite*

The L&SWR and the District Line

Wandsworth Town is the first station out of Clapham Junction on the L&SWR's Windsor line (see map). Opening in 1863, its entrance had an attractive pediment of brick, steel and glass but has since been replaced by a bland glass booking hall (Fig. 4). Next, on the L&SWR main line, Earlsfield station once had Jacobean ornamentation similar to Wimbledon Park but this too has been rebuilt in glass (Fig. 5).

Durnsford Road gave its name to the L&SWR power station that opened in 1915 and was enlarged in phases. It replaced an earlier power station at the end of the Waterloo & City line that was underground with coal and ash wagons using the Waterloo carriage hoist.

Fig. 6 shows class LN No. 30863 *Lord Rodney* passing Durnsford Road with a down express. Between the EMU sheds and the main line is the ramp for an electric loco to propel coal wagons to bunker level. Nearby is Wimbledon Park station on the District Line (Fig. 7). This humble station handles a zillion visitors every July for the tennis. Originally joint L&SWR/Metropolitan District Railway, the line opened in 1889 and was electrified in 1905. The square booking office has a nice touch of the Jacobean.

Wimbledon station was rebuilt by the SR in art deco style, looking just as smart and 'modern' to this day (Fig. 8). The approach is spacious and welcoming. Also surviving but disused is the Odeon style 'A' signal box seen in Fig. 9 with class 4MT No. 75079 charging through on a down express. Just departing for Earl's Court and Edgware Road is a District Line train including four clerestories. In 1963, Wimbledon goods yard was still active.

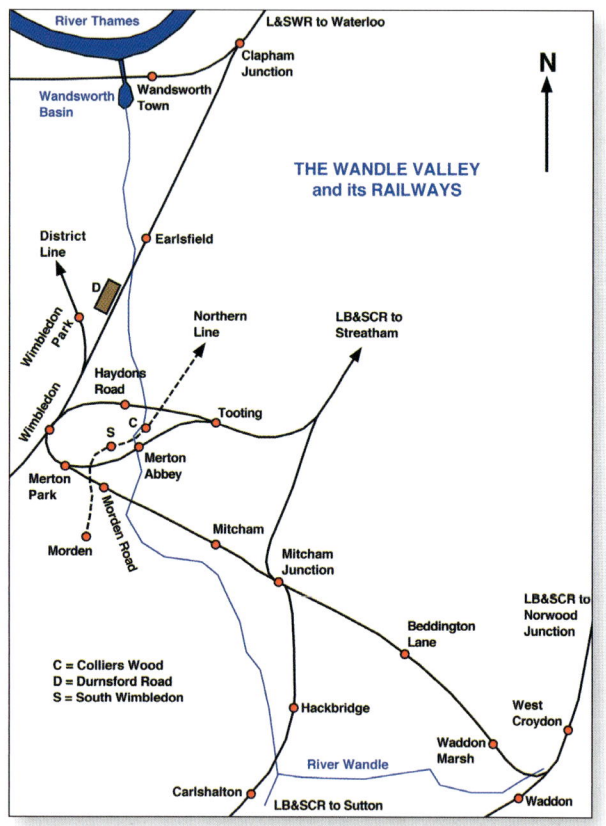

Fig. 4 *Alan Postlethwaite*

The Railways of the Wandle Valley

Fig. 5 *Alan Postlethwaite*
Fig. 6 *Colin Hogg/Bluebell Railway Archives*

Fig. 7 *Alan Postlethwaite*

Fig. 8 *Alan Postlethwaite*

![Fig. 9]

Above: **Fig. 9** *Alan Postlethwaite/Bluebell Railway Photographic Archives*

Right: **Fig. 10** *John J Smith/Bluebell Railway Photographic Archives*

Joint L&SWR/LB&SCR Lines

With reference to the map, the lines through Haydons Road and Merton Abbey were jointly owned by the L&SWR and LB&SCR, opening in 1868. The Haydons Road section was used not only for suburban services but also for specials terminating in London on the Brighton or Chatham systems. Boat trains used it, for example, on services between Victoria and Southampton Docks while Fig. 10 shows class T9 No. 30724 heading a special through Haydons Road, returning from Amesbury to Loughborough Junction. Fig. 11 shows the rebuilt up booking office at Haydons Road and some original valancing on the down canopy.

The original junction station at Tooting had four platforms. Fig. 12 shows the remains of the northernmost platform with the station house that had become a private residence. This platform served the up line via Haydons Road. This station closed in 1894, superseded by a new Tooting station, visible just beyond the road bridge. The junction to Merton Abbey, this side of the bridge, was severed in 1934. A modest goods yard to the right was then served from the south until the Merton Abbey line closed completely in 1975.

I walked the Merton Abbey line on a chilly, murky Sunday afternoon in October 1963 when all was eerily quiet and deserted. What I discovered in just 2 miles were three districts of quite different character – suburban in the north – semi-rural in the south – and industrial in the middle. Passenger traffic once included L&SWR push-pull services between Wimbledon and Moorgate and longer trains from Ludgate Hill running in a loop through Wimbledon via Haydons Road and Merton Abbey or vice versa. Merton Abbey closed to passenger traffic in 1929 when eclipsed by the Northern Line extension to Morden.

It was industrial traffic that kept the line open. Fig. 13 is a general view of the Merton Abbey industrial estate, while Fig. 14 shows the disused station. The original medieval mill of Merton Friary was water-powered from the Wandle. In the electrified era

Fig. 12 *Alan Postlethwaite/Bluebell Railway Photographic Archives*

Fig. 11 *Alan Postlethwaite*

of the 1960s, calico was manufactured here, used for both clothing and printing. The Regent Street fashion house Liberty printed textiles here. Other works were Shear's Copper Mills, Eyre Smelting Works and New Merton Board Mills. Finally came Lines Brothers, who manufactured Triang toys. All have now gone and the estate has become a craft village, a faint echo of the intense industry that once exploited the Wandle.

Fig. 13 *E N Montague collection/Merton Historic Society*

Fig. 14 *John J Smith/Bluebell Railway Photographic Archives*
Fig. 16 *Alan Postlethwaite*

Fig. 15 *Alan Postlethwaite*
Fig. 17 *Alan Postlethwaite*

The Northern Line Extension

Opening in 1926, the art deco of the Northern Line Extension is stunning to this day, as seen at Colliers Wood and South Wimbledon (Figs 15, 16). Morden terminus, however, has had its grand booking office and shops altered by the addition of a blue office block on top (Fig. 17). The booking office, however, retains its octagonal roof light (Fig. 18).

Morden is a surface station with five platforms under a steel and glass roof (Fig. 19). The green and white building contains offices, staff facilities and, until 1962, a control cabin. Beyond the station is the depot with workshops and thirty-two storage sidings, photographed here from a long footbridge (Fig. 20).

Right: **Fig. 18** *Alan Postlethwaite*

The Railways of the Wandle Valley

Fig. 19 *Alan Postlethwaite*
Fig. 20 *Alan Postlethwaite*

We regret space constraints alone have caused to split Alan's article into two sections. Part 2 will appear in SW54 and cover the Wimbledon to Croydon line together with biographic details.

(Some) USA tanks
Before the Southern

We recently came across a small collection of views of USA tanks seemingly in store 'somewhere in England'. Possibly this was at the supplies dump at Newbury racecourse as we know some at least of the engines purchased by the Southern had arrived from store here travelling by means of Whitchurch and Sutton Scotney to reach Eastleigh.

A bit of detective work and it was possible to identify five engines seen here in WD/Transportation Corps branding that were later renumbered when acquired by the Southern. Clearly medium to long-term storage was envisaged as covers have been provided to the domes and chimneys and the side of the cabs. With each also having their rods removed it is possible they had simply been worked here from previous locations. (Does anyone know where these particular engines had served between construction in 1942–43 and their acquisition by the Southern Railway?)

WD No. 1264 leads a line-up of stored engines. Light section track tends to imply wherever this was it is most likely a temporary facility. This engine was Porter built as their No. 7420, later SR No. 61 and BR 30061, and lasted until March 1967.

(Some) USA Tanks

Slight variation with shaded lettering on the tank side on No. 1277. One thing we can be certain of is the rods have been removed and the engine(s) then shunted into the positions seen – note, the crankpins are not aligned. Wooden packing has also been placed around the area of the crosshead.

A further variation, with No. 1968 in wartime livery: she would later be SR No. 65, BR No. 30065 and finally DS237. This example would also survive in preservation.

WD No. 1279, later SR No. 66 and BR No. 30066. This had started as Vulcan-built No. 4377 from 1942 and was withdrawn in August 1965. Was there ever a record kept of miles run both in the hands of the WD and later the SR? Notice, too, the limited-capacity coal bunker with a straight side to the rear.

Finally No. 1974, SR No. 73, BR No. 30073, in service until December 1966.

War Comes to Eastleigh

Aerial view of the locomotive works (centre) and carriage work (top right). The houses left bottom are those in Campbell Road with the locomotive shed out of camera. The view is north towards Allbrook and Winchester.

Prior to the March 2020 lockdown, your editor had spent some time at the National Archives researching Southern Railway and Southern Regions papers for the special issue on Oil Burning – SW Special No. 17. At the same time he happened to spy upon this item (Rail 648/19), which was duly copied and is now presented as something a little different and yet most interesting as representative of the time. The railway at Eastleigh was lucky, indeed apart from one occasion it could almost have been said to have led a charmed life. How it had prepared for totally different circumstances makes for an interesting read.

Eastleigh works – air raid precautions

As was the case with all the Southern Railway Company's Workshops, Air Raid Precautions were first considered in the early part of 1938, and the whole question was gone into by the Mechanical Engineer (Mr. E.A.W Turbett). The provision of suitable shelter accommodation for the employees was one of the major questions and proposals were submitted. These provided for the use of all available sites, including Stores Warehouse, Office Corridors etc. wherein the office staffs would be afforded certain blast protection. It became obvious that a larger scheme of shelter would have to be provided for the Workshop employees and eventually concrete slabbing trenches were placed around the outskirts of both the Locomotive and Carriage and Wagon Works.

The protection of essential buildings was also of consideration and such places as the two Power Houses had to be subject to blast protection. In view of the fact that the whole of the Eastleigh Railway area was one unit, (so to speak), it seemed certain that any Air Raid Precautions scheme would have to be considered jointly with the departments concerned, i.e. Locomotive Running, Traffic, Civil Engineers (including O.D.M. & Signal & Telegraph), Stores, Accountants and Accountants Statistical Departments.

The first point to decide was the responsibility for carrying duties of Chief Warden. Generally speaking, this duty had been placed upon the Station Master at other places on the system but in view of the peculiar circumstances at Eastleigh, the total area of Company's property being 220 acres, it seemed to the Traffic Department that the responsibilities would be too great to carry and it was eventually agreed that this should be done by a member of the C.M.E. Department. In the first instance Mr. F. Munns, Loco Works Manager, accepted the position of Chief Warden, Eastleigh S.R., and as he lived near the Carriage Works (with the greatest fire risk), this seemed a good arrangement. Subsequently, however, upon his transfer to Ashford, Mr. C.A. Shepherd, Carriage Works Manager, became Chief Warden. The responsibility for the formation of the A.R.P. scheme generally, and especially as it affected the staff and squad workers was one for the Chief Mechanical Engineer's Department.

The whole of the Railway Area was split into sections and it was decided that certain areas would appear on the plan for action with members of the various departments acting as 'Area' Wardens. This scheme provided as follows:

Locomotive, Carriage & Wagon Works – Chief Warden

Locomotive Works – Deputy Warden

Locomotive Works – West side of Straight – Area Warden

Locomotive Works – East side of Straight – Area Warden

Carriage & Wagon Works – Deputy Warden

Carriage & Wagon Works – Deputy Warden

Carriage & Wagon Works – West side – Area Warden

Carriage & Wagon Works – East side – Area Warden

Running Shed – Area Warden

Station – Area Warden

East Yard – Area Warden

Marshalling Yard – Area Warden

Engineers & Dutton Lane – Area Warden

Each department other than C.M.E. was asked to divide their own area into sections and suggestions made to them regarding the provision of Wardens to cover these sections. Obviously any scheme formulated had to provide for assistance with the Local Authority and several meetings took place with the representatives of that body which included their Engineer and Medical Officer of Health. They were particularly interested in the question of gas cleansing, first aid, demolition and rescue work.

It was agreed with them that there should be a mutual arrangement whereby the Railway Company and local authorities would do all in their power to help and provide assistance to each other. So far as this help was concerned, nothing could be expected from the Locomotive Running or Traffic Departments in view of the heavy demands that were expected in the way of war traffic. It was, however, thought that the Engineers Department would be prepared to assist with staff stationed at Eastleigh but similar difficulties were likely to be experienced in that Department also. The whole burden, therefore, fell upon the employees engaged in the two Main Works which included members of the Stores Department. Certain of the Accountant's staff were allowed to assist during working hours.

No A.R.P. scheme could be complete without an efficient Fire Service and to commence with it was decided to increase the permanent personnel of eight men to ten with the duties of Chief Officer (Works Fire Brigade) made full-time instead of part-time.

The arrangements that applied generally from the outbreak of war are therefore set out under their various headings rather than in the order of actual sequence to avoid as far as possible a disjointed account. Appeals were made from time to time for volunteers and in 1939 two-day courses were arranged at the Railway Institute, Eastleigh, for those desirous of obtaining the necessary information regarding A.R.P. Subsequently, appeals were made for employees to volunteer for squad work in first aid, auxiliary fire service, decontamination, demolition and rescue etc. and a good number offered their services.

As the success of any A.R.P. arrangements depended so much upon the goodwill of all the employees, the question of their participating in discussions affecting shelters,

The Chief Warden and Control Room staff. (Names are not confirmed but the position of Chief Warden was originally that of the Locomotive Works Manager Mr Munns, although he was later replaced in that role by Mr Shepherd, the Carriage Works Manager.)

facilities etc. was gone into and subsequently their representatives were invited to form an A.R.P. Committee with the Company's representatives.

This proved to be a very satisfactory arrangement and great help was given by all concerned on that Committee during the period of hostilities. All manner of questions were discussed and some of these appear in the resumé under that heading.

Control

The focal point in all messages was the telephone exchange which was situated at the Station. Suitable instruction was given to the operators regarding the dissemination of warning and other messages. Each department selected a point to which such messages were to be sent and instructions given to all concerned that these were not to be passed on to the public. Under the A.R.P. scheme it was agreed that the Control for the area should be the Gateman's Cabin at the Carriage and Wagon Works with the Gateman's Cabin at the Locomotive Works as the sub-control for the area west of the Portsmouth Line. This included the Running Shed etc.

A small arm carrying coloured discs was fitted in the Control for the use of the Chief Warden of any warning in operation. Originally these were as follows:

Yellow – Preliminary Caution

Red – Action Warning

Green – Raiders passed

White – Cancel Preliminary Caution

Having agreed upon a division of the Railway Property into areas the various Area Wardens posts were set up as follows:

Works
  Loco. No. 1 – Outside Pattern Shop
  Loco No. 2 – Outside Boiler shop
  C&W No. 3 – Inside Repair Shop
  C&W No. 4 – Inside Wagon Shop

Other Departments
  Running Shed – Superintendent's Office
  Outdoor Machinery – Inspector's Office
  Station – Station Master's Office
  Engineers – General Offices and Dutton Lane Depot.

During a warning each Area Warden had an assistant and messenger and in the event of an incident it was his duty to patrol the area for which he was responsible and report back to main or sub-control. It was the duty of these controls to regulate the flow of assistance required. Each of the Ambulance Rooms was linked by automatic telephone to control but as telephones might not have been in Working order, they were not to be relied upon and arrangements for messengers made.

It was agreed with the Local Authorities that any calls for assistance must emanate from the Works Main Control and that no help would sent unless this was done. This was necessary to avoid unnecessary calls being made upon them.

Employees' representatives from the A.R.P. committee.

An additional night duty Gateman was provided at each Works to act as messenger or for any other work that might be necessary. The duties of A.R.P. Control were carried out by clerical staff during the day time. No special provision was made for a 24-hour service at the outset of the war apart from personnel being 'on call' in the event of a warning. There was not the male clerical staff available for a 24-hour shift working to be provided specially.

The Chief Officer of the Fire Brigade was on continuous night duty for a considerable time and materially assisted in matters affecting the control. Anything untoward was immediately reported to the Chief or Deputy Chief Warden. With the introduction of the Works Warnings within the Town Alert, a night shift of employees' representatives was introduced and this is further explained in the' appropriate section.

The arrangements laid down regarding the necessary control of the areas remained until the end and so far as can be judged, worked as satisfactorily as could be expected under the trying circumstances.

Local authorities' schemes

As with other Railway Towns it was certain that the Local Authority had to rely to a large extent upon Railway workers to assist in any scheme they formulated. It was therefore agreed in the first instance that all those employees enrolled in the Scheme should be allowed to leave the Works when the Town sirens were sounded. The number involved in the early days was in the region of 150 from the Works and these had to get to their appropriate assembly point in the Town. When the internal warning system was introduced, however, the numbers leaving on the Town warning was reduced considerably and consisted mainly of Key Wardens, etc. A number of the remainder took up their positions when the Works warnings sounded. It was agreed that the Works Fire Circuit Siren should be used as a warning to the Townspeople and was disconnected from the Works Fire Circuit for this purpose. A number of tests were carried out periodically. When it became obvious that all the sirens in the town should be synchronised owing to warning messages not being received at the various control points at the same time, it was agreed that a relay should be inserted and connected with the town's control. Prior to this the siren had been sounded by a responsible person in the Main Control at the Carriage Works. At the outset of the war it was agreed that the facilities in the Town for gas cleansing, first aid casualties etc. should be at the disposal of the Railway Company should such necessity arise and Gas Cleansing Centres and Dressing Stations were provided with this in mind. A 24-hour service was always available and agreed that priority was with industrial plants.

A.R.P. Traning

In 1938 certain of the staff of all departments were required to be trained in Air Raid Precautions and this necessitated visits to the Brunswick Institute in London to those selected for the purpose. It became obvious that as many of the staff as possible should receive instruction in order that they would have some knowledge of what war meant and to become 'precaution' minded. These early instructions consisted mainly of anti-gas precautions and had no connection with the training that was to follow to the volunteers for the various sections of the civil defence squads. As mentioned previously, two-day courses were eventually arranged for all the employees who had volunteered for same and these were held in the Railway Institute, Eastleigh. Each member who satisfied the Instructor that he had absorbed all the necessary knowledge was issued with an A.R.P. badge. From these employees, volunteers were sought from which to form the various squads required in the Works, and squads of workers as follows were formed:

Fire-Fighting
Fire-Watching
First Aid
Demolition/Rescue
Decontamination.

A number of lectures was arranged in the Works and the Company's A.R.P. instructors visited the premises to carry these out by means of the A.R.P. Training Coach and at other times on the premises.

The fire-fighting training was under the jurisdiction of the Chief Officer of the Works Brigade and included drills of all types, running out hoses, use of stirrup pumps, hose carts, and finally the Fire Pumps.

Training in First Aid took place after normal working hours, often in conjunction with the Fire Services, when realistic incidents, including incendiary bombs, were staged. This proved of value to all personnel taking part and was greatly appreciated.

Works A.R.P. committee

This Committee, formed to discuss all questions affecting the safety of the employees, consisted of the following:

Asst. Mechanical Engineer or Works Superintendent
Loco Works Manager
C&W. Works Manager
Representatives from Loco Works Employees
Representatives from C&W Works Employees
Representatives from Clerical Staff
Representatives from Supervisory
A. R. P. Staff organiser.

On the 16 September 1940 the first question raised was the length of time that was being wasted as the result of the Town Alert warnings, where no enemy action took place. Meetings of the employees were addressed by their representatives but their decision was in favour of continuing the system of Town Alerts. In October 1940 the question of warning within the alert was again discussed and following visits to Gun Sites and other firms it was decided by the Committee to give the information obtained from this source a trial. Although no idea was given of the methods adopted by the Military Authorities it was subsequently learned that it was the result of Radiolocation.

Many other items dealt with by this Committee included:

Warning lights throughout the Factory
Warnings close to or during meal hours
Lighting of Shops
Company's A.R.P. Personnel
Night shift. Workers safety
Trenches and other Shelters
Control Room Staff
Methods of warnings
Electricity supply failures
Personal Injuries
Payment of staff late for duty
Staff injured on way to work
Gas Exercise
Spotters.

Works warnings

At the outset these were given by means of the Fire Alarm siren situated in the Carriage Works. Subsequently each of the Workshops and offices were fitted and controlled from the Main Control Room set up in the Carriage Works.

During the latter part of 1940 it was not uncommon to spend two or three hours in the shelters as the results of the Town Alert and in November of that year agreement was reached upon the internal warning in the Factory, within the Town Alert. The Control Room as set up in the Carriage

Works was in direct contact with a Gun Site and from information passed regarding Friendly or Hostile Aircraft, plotting was done upon a chart shewing the actual position and direction of aircraft at any time day or night. It was agreed that a 25 mile radius be worked to and when Hostile of doubtful aircraft were approaching at this distance the Works siren was sounded.

Similarly when aircraft passed out of this radius the All Clear was given. This system proved to be a great time-saver and soon after its adoption was appreciated by all concerned. Representatives of the employees were called upon to take part in the duties of Control of Warnings and eventually they were employed upon a 24-hour basis for the purpose.

Wardens

In view of the numbers of trenches and shelters it was certain that a large number of Wardens would be required and it was decided to draw these mainly from the Supervisors and Officials. Each received instruction and were given demonstrations by Government-trained Railway Instructors. One Warden, with a deputy from the employees, was detailed to each trench and lists of employees using each were furnished. In the early days a roll call was decided upon should warnings occur, but after a time this proved unnecessary and the boards provided at the shelters were marked with the total number only of those taking cover. The duties covered those normally expected of Wardens and following any enemy action it was their duty to report such incidents as might be necessary to the Area Wardens. The area Wardens passed on any reports to Control for action.

Fire-fighting

Of all the A.R.P. Services this section was of paramount importance. From the outset it was realised that fire was the most likely danger from bombing. It was decided in 1939 that there should be a Fire Pump available in both Works in case of breakdown of water mains. The pump at the Fire Station controlling the supply of water from the river to the Works was suitably protected. As the result of Air Raids during 1940/41 it was decided to increase the number of Fire Pumps to six, four of which were placed in the Carriage Works and two in the Locomotive Works, housed in suitably protected shelters. In addition to this, two were provided for the use of the Running Department to be used in case of failure of water supplies at stations. Additional fire equipment in the shape of Foam Branch Pipes, Foam Fire-Engine, Breathing Apparatus was also provided. In 1939 volunteers to the Fire Service were called for and 160 employees volunteered, for duty either as Fighters or Watchers. Squads of six men were formed into Fire-Fighters and were equipped with helmets, respirators, rubber boots and axes. It was essential that this service should be on a 24-hour basis. A good number of the volunteers agreed that they would report to the Works after normal working hours in case of emergency.

The number of permanent Brigade Firemen was increased from eight to 25 in the early part of 1941 and arrangements made for the majority to be available on night duty. These men were also acting as Works Watchmen, patrolling the whole of the Eastleigh Area. During the war period all the fire-fighting arrangements for each of the departments, Motive Power, Traffic, Engineers and C.M.E., came under the jurisdiction of the Works Fire Brigade Captain. He was responsible for training all members of the staff of all departments who had volunteered for such service. The upkeep of Fire Fighting equipment was also one for the C.M.E. Department.

From time to time contacts were made with the N.F.S. (National Fire Service) and their members often visited the Works to obtain some idea of the layout in case of being called in, in an emergency. It was realised by them that railways presented a problem much different to that expected in other factories and that it would be necessary for the Works Brigade to materially assist them on all occasions.

First-aid

This was considered a most essential section of the services and the Ambulance Rooms in both Works were protected from blast by means of sand bags erected round the entrances and walls of the rooms. From the employees who volunteered for this service, a number were allocated to trenches in charge of the first-aid box provided for emergency and one other who reported to the Ambulance Room for duty following any incident in the Works. Arrangements were also made to have a squad of men always available at the Ambulance Rooms during an alert and for an additional Ambulance Room Attendant as well. The Ambulance Room Attendant acted as the Warden for this section and was responsible, in each of the Works, for about 40 squad workers.

Holding the responsibility of first-aid for the whole of the Eastleigh Area, the C.M.E. Department had again to call upon the services of its personnel to be available after normal working hours and a good number of employees agreed to report for first-aid duty in emergency. All the volunteers in this section held St. John Ambulance awards in addition to their A.R.P. badges.

Demolition/rescue

As with the other services, volunteers from the staff were formed into working squads under the direction of a Warden. Provision was made for the staff to assemble at given points in the Works where suitable protection was given by means of sandbags etc. Hand carts containing tools, such as picks, shovels, etc. etc. were made available. Attached to the squads was an employee trained in the use of oxyacetylene cutting. Advantage was taken of the County A.R.P. Training scheme and two members were given a week's course at Basingstoke.

Decontamination

This side of A.R.P. commenced with the training of volunteers and a number of squads of six men each were formed. They received instruction in the Works and followed up with an extensive two-day course at Southampton. This was a very pleasant job to look forward to but all concerned entered into the spirit of the business. It became necessary to consider the decontamination personnel and staff and after careful thought it was agreed the Time Office in the Locomotive Works could be converted into a cleansing centre. This was proceeded with and when completed formed a suitable building for cleansing purposes.

This arrangement however was not sufficient to cater for all the employees in the Works and further consideration had to be given to the question of cleansing facilities for female staff when they were employed. It was decided eventually to provide additional facilities and arrangements made for showers in each of the Works Lavatories used by the Workshop Staff. Additional facilities consisted of showers etc. for the female employees.

Suitable overalls, plimsolls, etc. were provided for the use of the squad workers in the event of their own clothing becoming contaminated.

In view of the fact that the local authority might not have been in a position to cleanse all the protective clothing of various departments, consideration had to be given to the question of a suitable laundry for the purpose. It was eventually agreed to install this in the Locomotive Works.

Appeals were made to the other departments at Eastleigh to provide volunteers for this work and four squads of women from the Works and Station were formed and given the necessary instruction. The laundry provided for the use of the available boiler and the boilermen were placed on call for a day and a night shift. Suitable training was given them.

Gas detection

Gas detection boards were placed at various points within the area and special gas detection paper for use with hand sticks provided for the use of Wardens.

Amongst the essential equipment was the rattle and whistle for use of Wardens and also the Warning Bell (one in each Works).

Gas Security officers were selected under the guidance of a member of the Laboratory Staff and exercises in conjunction with the Local Gas Identification Officer carried out.

Emergency breakdown gangs

The Loco Running Department did not expect to be able to handle all incidents without considerable assistance and with this in view the C.M.E. Department were asked to provide four gangs of twenty men each who could be called upon in an emergency. They were formed of employees engaged upon Construction and Repair of Carriages and Wagons. It was necessary for them to hold certificates of occupation and to carry identification cards containing their photograph. This was essential in order that they could visit a prohibited area without too much difficulty.

Blackout

This was a problem that was discussed in 1938, it being certain that if war did break out this was one of the first things that would have to be done at short notice. Decisions were made regarding restricted lighting in workshops and yards, blacking out of windows etc. It was agreed that blackout curtains must be one of the provisions and arrangements agreed to this end. All the glass fitted to the roofs of the various workshops had to be removed shortly after the outbreak of war and was replaced by felt: the area of glass in the roofs was approximately 50,000 sq.ft. This had the effect of keeping out a great deal of the normal daylight and the use of artificial light upon a large scale became necessary. Working under such conditions in the factory was far from ideal and arrangements were made to provide the best lighting possible under the restricted conditions. A tremendous amount of time had to be spent, in addition to all the necessary material used, in conforming to the various regulations laid down from time to time. Moveable shutters were fitted in the roofs, where possible, to afford some relief from the blackout.

Shelters for staff

In September 1938 discussions regarding trench shelters took place and arrangements made for provision on the outskirts of both Works. These were of the concrete slabbing type and contained shovels, picks, hand lamps, chemical closets etc. Use was also made of ground floor Office Corridors, Stores Warehouses etc. From time to time additional single shelters were provided in some of the Workshops for individuals who chose to remain close to their work during warnings, but this was not encouraged to any great extent by the employees' representatives in view of the risks involved. Each of the shelters were numbered and those on the outskirts of the Works provided with a light for use during warnings.

Towards the end of 1943, when it appeared certain that invasion of the Continent was to take place it was decided, in view of the increase of staff, to provide additional shelters and agreement reached that they should be of the surface type. It was only necessary for these to be erected in the Locomotive Works.

The various grades of employees in the Shops were given Shelter Numbers so that there was no congregation of too many of each grade in one shelter. In other words 'dispersal' of trades in the event of warnings was aimed at.

Arrangements after normal working hours

In the first few months following the outbreak of war it was felt that certain of the Wardens should be asked to volunteer to visit the Works when a warning was sounded and a suitable roster was drawn up provided for two Wardens to visit each Works. Eventually it became obvious that one in each Works would cover the emergency and with the Chief Warden and Control Staff reporting at the same time, a fresh roster was made. As a general principle the Foremen and other supervisors living in the area performed this very essential duty.

It was considered necessary for numbers of the various A.R.P. squads to be asked to volunteer for A.R.P. duties after working hours and it is interesting to record that a great number did so, especially in the Fire Service. With the frequency of warnings and their long duration, it was eventually decided to restrict the numbers reporting for duty each night, in the squad's own interest, because of loss of sleep.

Rosters were prepared and arrangements made for the Fire Service Personnel to report to the Works upon alternate nights. Similar arrangements were made with the Demolition/Rescue and First Aid Squads.

Fire guards

The first Fire Prevention (Business Premises) Order was issued in January 1941 and in prescribed areas called for sufficient personnel to be always available to discharge fire prevention duties at all times on the premises. With the Fire Brigade and Auxiliary Fire Service staff reporting for duties during Town Alerts it was considered that the Works arrangements were satisfactory.

With the issue of Fire Prevention (Business Premises) (No. 2) Order the position had again to be reviewed but as the works were open for 61 hours per week the numbers of employees likely to be available would be small. In September 1941 the preparation of rosters was gone into based upon the possibility of a 56 hour week; and the question of suitable amenities was discussed with the employees' representatives.

The necessary facilities were provided as follows:

Sleeping Accommodation:
Loco Works:
    Time Office and two old coach bodies.
    Top floor of Main Offices
Carriage & Wagon Works:
    Water Tower and three old coach bodies.
Washing arrangements:
Loco Works
    Time Office and General Offices
Carriage & Wagon Works:
    Water Tower and General Offices
Canteen:
    Loco Works and also Carriage & Wagon Works: open from 6p.m. to 10pm.

The number of Fire Guard parties to be provided was 42, of three persons each. In view of the number of Fire Brigade full-time staff on duty it was agreed that in the Carriage a Wagon Works the number of parties should be 20 and in the Locomotive Works 15. With the increase in the night shift, however, it was possible to dispense with compulsory Fire Guards with the exception of Saturday and holiday periods in the Locomotive Works.

One of the largest difficulties of the scheme was the provision that no one should be called upon to perform Fire Guard duties if working 60 hours or more each week. This meant the keeping of proper records of hours worked by all compulsory Fire Guards. As each member of the staff was liable to be called upon by the Local Authorities, certificates of exemption had to be issued every three months either to cover the duties performed on the Business Premises, or to cover the fact that 60 hours per week were being consistently worked.

A Fire Prevention Committee comprised of employees and employer's representatives was set up and methods of rostering agreed upon. Daily rosters of personnel were posted in each of the Works and in view of the fact that a good number of the staff took advantage of the fact of the Works being open for 61 hours each week to claim exemption on these grounds it was not possible to roster the required numbers under the orders.

The Fire Prevention scheme had to-be approved by the Ministry of Transport's Regional Fire Prevention Officer and several visits were paid to the Works by him. The scheme provided for the use of employees as Fire Spotters and for some time this was covered by the Full-Time Brigade. During alerts, service in this direction was also rendered by the Gatemen.

Each of the sleeping posts was connected to the system so that the Fire Guards on duty were only summoned when the Works warning sounded, mainly within Town Alert.

The training of the personnel was undertaken by the Company's A.R.P. instructors with the practical side of Fire being covered by the Fire Brigade captain.

A scheme of Fire Watching points was evolved to which Fire Guards reported upon the sounding of the Works Warnings and were placed upon the outskirts of the works near fire-fighting equipment. With the introduction of the No. 3 Order the whole of the Railway Area was treated as one Sector and in accordance with the arrangements of the Sector Plan was divided into various blocks. A direct telephone line with the N.F.S. and the Locomotive and Carriage Works was installed to ensure assistance being obtained with the minimum of delay.

Air raids

The first warning to be received at the Works was the Preliminary Yellow on 5 September 1939 at 10.7am. followed by White a few minutes later. As is well known, activity on the part of the enemy so far as aircraft is concerned did not develop seriously until 1940 and the first Red, or alert, warning took place on 7th June 1940 at 12.55a.m. and was in operation for two hours. During September and October of that year the

Bomb damage; the motor and cycle sheds.

warnings were of daily occurrence and a good deal of time was spent in the works shelters as a result.

The first damage from aircraft occurred on Tuesday, 8 October, when low flying machines travelling in a south-easterly direction passed over Eastleigh and dropped bombs, some of which fell upon the Locomotive Running Sheds, within a stones' throw of the Works. The Running Shed office was struck, two of the staff being killed and three injured. Calls for assistance were made and members of the Works and Shed Rescue squads worked well to reach the injured persons. The Company's property received no further damage until the night of Sunday 19 January 1941 when during an alert lasting four hours, bombs were dropped on both the Locomotive and Carriage & Wagon Works. The warning was sounded at 7.9p.m. and aircraft were soon passing overhead in a northerly direction. At approximately 8.45p.m. 16 bombs were dropped in the area, as follows:

- Two on permanent way north of East Signal Box
- Two on permanent way near South Coast Carriers
- One near cycle sheds in Carriage & Wagon Works
- One near Paint Shop in Carriage & Wagon Works
- One near Boiler House in Carriage & Wagon Works
- One between Brake Shop and Smithy in Carriage & Wagon Works
- One on Lifting Shop Roof in Carriage & Wagon Works
- One in Yard in Carriage & Wagon Works
- One unexploded in Traverser Road, Carriage & Wagon Works
- Two on permanent way South of Transfer Shed

Cycle Sheds etc. A large portion of the Cycle Shed and Motor Car Sheds was demolished by the bomb which fell near these Sheds. This also shattered all the windows on the North side and blew down a number of doors in the Carriage Works General Offices, Stores and Accountant's' Offices. The outside staircase to the Accountants' Offices was also damaged. The petrol pump at the Stores gate fired and was rendered useless.

Paint Shop. The north-west corner of the Paint Shop received damage. A piece of brickwork approximately 10'0" x 4'0" was blown down at ground level. Many windows along the wall and glass in the roof was shattered. A gas main and a water main were broken at this point, also the Works clock and clock tower received damage. The windows of the Control Room and Time Office were shattered. All windows in the South side of the General Offices were shattered and a number of doors blown down. Damage was also done to the Ambulance Room.

Boiler House. The bomb which fell within a few feet of the Smiths' Shop Boiler House damaged the Works hydraulic main, compressed air main and secondary steam main. The coke storage bin was blown down and damage caused to the corrugated iron building housing the hydraulic machine etc. The automatic starter of the vacuum pump was damaged beyond repair, and also the starter of the 60 H.P. hydraulic motor received minor damage. The north-east corner of the Iron and Steel Stores received damage. A number of windows were shattered along the West wall of the Smiths' Shop and a quantity of glass in the roof broken. The wood and corrugated iron building housing the hydraulic ... *(the remainder of the text is missing – Ed.)*

Brake Shop &- Smithy. The timber-built lean-to sheds containing the Smiths' Shop blowers were blown in and the 25 H.P. motor and starter of one of them received damage. Eight pairs of doors at the North end of the Lifting Shop were damaged, 6 pairs badly, but the building fabric appears to have stood up to it. The timber and iron building containing the two boshes for cleaning axleboxes received damage to the iron sheets and windows. A quantity of glass was shattered in the roof of the Lifting Shop and a number of windows at the south-west corner of the Body Shop were blown in.

War Comes to Eastleigh

Bomb damage; north-west corner of the carriage paint store. This and the following images relate to the raid of 8.45pm, Tuesday, 8 October 1940.

Bomb damage; north-east corner, iron and steel store. Five people were killed in this raid.

Bomb damage; north end of brake shop.

Bomb damage; south-west corner of lifting shop.

Lifting Shop. This bomb fell in the south-west corner of the Lifting Shop and it appeared to explode on contact with the roof. A portion of the roof about 35'0" in length in one bay fell in, causing damage to Waterloo and City stock. A large quantity of glass was shattered in the Lifting Shop.

Yard. South of Lifting Shop.

This bomb fell outside the south-west corner of the Lifting Shop and landed in the centre of a new underframe. A hydraulic main was fractured by the explosion and the pipe broke a few days later.

Fires.
With regard to the fire at the Petrol pump mentioned previously, about 250 gallons of petrol were contained in this pump. The fire was extinguished within ten minutes by the Fire Service under the supervision of the Brigade Captain W.H. Cawte with the aid of Phomene extinguishers.

In the Lifting Shop one or two old coats were thrown to the roof and became ignited. This outbreak was quickly dealt with by the Brigade.

The fact that the Petrol Pump Fire was extinguished within ten minutes speaks well of the initiative shewn by all members of the Fire Service who rendered assistance under actual raid conditions.

The pump was within ten yards of the Stores warehouse and had the flames reached this point some considerable damage would have resulted.

At the same time as this fire was receiving attention five members of the Casualty Service who were on duty lost their lives as the result of the bomb dropping near the Paint Shop.

They were:

Ambulance Attendant R.G. Gillinghan

Tinsmith A. Godfrey

Bodymaker F.E. Ball

Labourer G.W. Henbest

Clerk L. Gaiger of the Stores Dept.

The sixth member of the party, Painter A.H. Welch, was absent from work for some time suffering from shock.

The Locomotive Works First Aid Party, Leader, R.H. Roberts, was called over to render assistance together with the Demolition/rescue squads from both Works. The services functioned in a praiseworthy manner upon this occasion and received a good deal of help from members of the Home Guard who were on duty. As it was not possible to decide under blackout conditions whether a number of unexploded bombs were also on the premises as a result of this raid the Works were closed for a short time on the following day. With the exception of one or two of the shops in the Carriage Works, the employees lost very little time as a result of damage and with 'all hands to the pumps' the repairs were under way and work proceeded with the minimum of delay.

During the alert of 12 February 1941 a bomb was dropped upon the Dutton Lane Recreation Ground. This was the only damage upon this occasion despite the fact that bombs were dropped upon other parts of the Town. On the 10 April 1945 an unexploded bomb was found at the Running Shed in a coal stack, presumably from the raid of Tuesday, 8 October 1940. This was removed without trouble.

War Comes to Eastleigh

Bomb damage; lifting shop. Clearly visible in this and the next view are Waterloo & City line rolling stock. This had been built at the Dick Kerr works at Preston and was delivered in batches through 1940, commencing tube work on 28 October. Their presence at Eastleigh must have been for checking and commissioning, although for these vehicles at least that process would be delayed.

Bomb damage; lifting shop.

It can be said with certainty that the Eastleigh area was extremely fortunate in not having received greater damage than actually occurred. Night after night aircraft passed overhead, the worst was anticipated, but the place was spared. During the raids upon Portsmouth and Southampton the bomb explosions could be seen and felt and during the fire raids on Southampton, in particular, everything seemed to be happening in the 'next street'. The Works A.R.P. staff who had agreed to report to the Works during raids carried out their duties in an admirable manner and under such a voluntary arrangement much could be said regarding the conditions of such service. The thanks of the Company is due to them for all their efforts.

As the Works was in close proximity to the coast it was to be expected that enemy aircraft would continually be in the neighbourhood and this is reflected in the number of alerts the were sounded locally. These totalled 1699. During the 'Battle of Britain' dog fights could be heard going on above the low lying clouds and bullets from machine gun fire rattled down on the roofs of the buildings.

Dispersal of stores

It is of interest to record that stores of all descriptions were dispersed to various stations of the system to avoid possible damage by air raids. Items such as Timber, Diesel Oil, Engine Fitting, Upholstery materials, Carpets, Wood Patterns etc. were stored at stations including Itchen Abbas, Micheldever, Wilton, Tongham, Farnham, Wherwell, Privett, and Fordingbridge. A total of £160,000 worth of valuable material was moved and stored in this manner from Eastleigh Works.

Invasion

Many preparations were made when it was thought the enemy was likely to consider the invasion of this country.

Suitable dockets were prepared for issue to the staff in the event of their being evacuated to other parts of the country at short notice.

So far as the Works were concerned, decisions were to be made regarding dismantling of essential equipment, such as Power House Plant. Certain details from Locomotives in Shops were to be removed and loaded into covered wagons set aside specially for the purpose.

We might conclude this piece with some slight humour. The late Harold Gasson, in his reminiscences book *Firing Days* based on his time as a fireman at Didcot, recounts the first time he was on an engine that arrived at Eastleigh shed; this would be around 1942. Being young and keen, Harold asked permission of the Foreman to look around, which was granted. Harold takes up the story, 'He (the Foreman) walked with me to the top of the Shed, and in our conversation he pointed to a sand bin in the corner of the Shed. It was kept there usually full of sand to deal with incendiary bombs. He told me that a few weeks previously he had been caught outside the shelter during a heavy raid, and he had dived into that sand bin for protection. The funny part about it was that he was on the large side and the bin was not all that big, and he had since tried to get into it again but without success; such is the incentive for self-preservation.'

Bomb damage; underframe outside lifting shop.

Bomb damage; entrance to shelter. At the peak it was not unknown for the workforce to have to spend two to three hours of their shift not working and in a shelter.

Down to Earth Part 3A
Ex-LBSCR Stock
Mike King

We will now consider ex-LBSCR stock. Both Brighton and Lancing Works appear to have discovered that selling old carriage bodies brought in revenue from an early date, since several examples from the 1850s and 1860s have survived to become the property of the Bluebell Railway and now reside in and around Horsted Keynes yard.

This really ancient specimen (actually two, as there was another identical one behind) was grounded at what the photographer describes as 'Friday Street' – a location giving a number of possibilities. A Google search reveals three: in Eastbourne, at Warnham (north of Horsham) and near Abinger in Surrey. The writer thinks he can recall these bodies (having been bought an ice cream here about 1960!) and feels this must have been the Abinger location, although one published source claims Eastbourne, however this is now an urban, rather than a countryside, location. The bodies themselves are from the Craven era – enclosed 22ft seconds or thirds from the early 1860s and probably grounded about 1900. In fact, there are three bodies, as the cut-down portion of another provides support below the 'shop' in the picture – the site itself is on a steep bank.
E R Lacey

By no means all will prove suitable for restoration to running condition and several have been acquired simply as a source of spare parts or to be restored as static exhibits, but a visit to the carriage and wagon department at Horsted (when things return to normal) is definitely recommended to view the Stroudley four- and six-wheeled carriages now either already restored or in the process of restoration. Likewise, the Mid-Hants Railway have several more under restoration (not on public view), while on the Isle of Wight, a number of later vehicles are already in passenger traffic. Apart from those on the Island, the others have been recovered mostly from Surrey and Sussex, with most of the Mid-Hants examples coming from Hayling Island – another noted coastal location for grounded bodies – not just of Southern origin either as several ex-GWR clerestory coaches also resided there.

Undoubtedly, more Brighton coaches remain somewhere in the countryside or in Sussex coast locations and the writer discovered a Stroudley brake third in a garden on the edge of

A Stroudley six-wheeler grounded in Polegate goods yard, seen in August 1938. This was originally an LBSCR Diagram 37 four-compartment first – one of about 150 built between 1871 and 1881. This example is one that had been altered to a brake coach, post-1900, with the right-hand end compartment converted into a guard's van, with double doors and end windows. Quite a number were so altered and some were downgraded to seconds or thirds at the same time. The one restored survivor, LBSCR No. 661, is an example of this type, finally ending up as brake third No. 1648. It has now been returned to a full first and is operational on the Bluebell Railway. It ended its days built into a bungalow at Bracklesham Bay in 1924 and was acquired by the Bluebell in 1983. It was allocated the SR number 3704 but this was never applied. The exact identity of the Polegate coach is not known for certain but could easily be one of the other four to survive the Grouping to be allocated SR Diagram 175 and SR numbers 3700–3. None were ever renumbered and all were withdrawn in 1923–24. A further eleven became SR stock as full firsts but only one, SR No. 7512, was renumbered and repainted into SR livery, lasting until 1929. Note the sagging roof – caused by the removal of the internal partition beneath. The composite coach body near Petworth noted in the text was an example of this design. *E R Lacey*

Ford airfield as recently as 1988 – it still showed traces of LBSCR lettering. This has now gone, presumably demolished sometime around the millennium. From time to time the Bluebell Railway is alerted to the existence of a carriage – usually when a site is redeveloped – and even if the whole coach is not recoverable, often many of the parts are, for use in restoration of vehicles already in their possession. Withdrawal of Stroudley coaches began before 1900 and the last mainland example went in 1929, but it is quite obvious that most sales occurred during this period. Of in excess of 100 Stroudley coaches allocated SR numbers in 1923, as many as forty are recorded as being sold over the following two years – and this would be in addition to all those sold by the LBSCR prior to Grouping. Vehicle bodies grounded by the railway actually seem less common, however that changed once withdrawal of bogie coaches began after 1930 – yet again illustrating the fact that shorter (i.e. 32ft and under) carriages remained the most popular to be built into dwellings.

Come the Second World War and many 'balloon' coaches were grounded – both on and off railway land – although this was not the first time that such vehicles had been sold off. Several had become ambulance cars during the First World War and a few of these subsequently 'landed' near the Grand Union canal between Greenford and Northolt, Middlesex, even though most of their underframes were reused at Lancing. Another feature of ex-LBSCR coaches was that many were rebuilt into electric stock from 1925 onwards and this led to spare portions of bodies being grounded at various locations – again both on and off railway property. As such, a three-compartment section of bodywork, with doors fitted at one end, could serve admirably as a garage for a car! For many years a three-compartment first-class section of a body could be seen as a garden summerhouse alongside the A285 road about a mile south of Petworth station, going towards Duncton. When finally examined, it yielded the number 347 on the inside of one door rail, leading to the conclusion that it was part of a four-compartment first that had been downgraded to a composite and withdrawn in 1922. Occasionally, a body might actually be split into two and both portions grounded at separate locations. Identifying these could be quite difficult and the New Cross Gate foreman painter's list could prove useful on these occasions.

While only the roof in the last picture might have been sagging, here we have a Stroudley 26ft four-compartment coach where just about everything is sagging and presumably the body was not long for this world! Photographed on a farm at Hayling Island on 24 June 1951, traces of LBSCR lettering, together with the class designations of second and third, appear on some of the doors. This coach might have started life either as a suburban first or as a main-line, second-class vehicle sometime between 1872 and 1888 – but more likely a second of which 120 were completed between 1876 and 1888. Most were downgraded to third class in 1911–13, leaving just eleven thirds (to SR Diagram 58) and four seconds (to SR Diagram 242) remaining at the Grouping. Most were withdrawn by 1925 without being renumbered except for two that were re-classed as composites and were sent to the Isle of Wight, where they lasted until 1931. With the eye of faith, the number 1539 might just be discerned in the waistline panels, meaning that the coach was scheduled to become SR No. 1817 and was recorded as being sold in October 1925, without being renumbered. It was last formed in SR set 974 – the final set composed entirely of Stroudley coaches on the mainland – and comprising no fewer than twenty vehicles (nineteen four-wheelers and a six-wheeled first). This coach is mounted on a cart, which was not unknown, allowing them to be moved around as necessary. Just to what use the vehicle had gravitated by the date of the picture may only be guessed at, but perhaps the goose in front knows better! *R C Riley/Transport Treasury*

Quite a number of four- and six-wheeled van bodies were also grounded – mainly on railway land – or for use as a garage somewhere, but the writer is only aware of one that still survives today and this is now some distance from the nearest railway and a long way from the former LBSCR. One other 'mystery' van body is also known. In 1994 the Bluebell Railway was alerted to the existence of an 'LBSCR' van body many miles from home at Churcham, Gloucestershire. This was duly visited and the vehicle had all the hallmarks of Brighton origin, but was of an unusual length for a van (32ft) and did not fit any known diagram. It has been suggested that it was one of two built specifically for ambulance train use in the First World War, but remains unidentified. Even then, it was in very poor condition and as far as is known it has now been demolished.

Several Pullman cars – once very much a part of the 'Brighton' scene – were also grounded. Two (*Devonshire* and *Albert Edward*) resided for a long time within Preston Park works yard, while two more were retained on their wheels as office accommodation and storage at Lancing, while others were sold off to be incorporated into homes in Sussex; especially around Lancing. Getting these to their final locations must have proved quite a challenge, as they were a good deal longer and heavier than most railway-owned carriages of the period – although probably providing somewhat superior accommodation.

We will now take a look at some more individual examples. As previously, carriage numbers and dates will be quoted, if known.

Groundings by the railway themselves seem less common, but here we have a Stroudley 26ft five-compartment third doing duty as a shunters' cabin at Eardley Road carriage sidings, Streatham, and photographed on 1 July 1951. It has been 'improved' by the addition of a corrugated iron roof and a shiplap timber extension at the far end – a store or the privy maybe? The stove is also prominent, but note the horseshoe hanging on the end of the coach. Perhaps this was to give luck and ensure no accident befell the occupants during their work. The coach was LBSCR No. 229, built in 1885 and withdrawn in November 1921 and one of over 600 such coaches completed between 1872 and 1891 – the most numerous type built by the LBSCR. Of these, just forty-one were allocated SR Diagram 57 at Grouping, with two more as second class Diagram 241 and, apart from two sent to the Isle of Wight and several more incorporated into the Lancing Works staff train, all had been withdrawn by the end of 1925. Several other Stroudley vehicles of this type are noted in Brighton carriage registers as being grounded at various other carriage sidings, several others went to 'Newhaven Seaplane Station', while two were sold to Major Fowler for use at Seaford Camp and another to Lt Col Perkins of Bognor. One wonders to what use he put his vehicle? *D Cullum*

We now move to Billinton six-wheelers. At Selsey West Beach on 17 May 1948, we see composite SR No. 5856 (ex-LBSCR No. 326) in the process of demolition. Another identical vehicle stands behind (SR No. 5844), while at least one more (SR No. 5826) was also present nearby. All were formerly 32ft five-compartment seconds dating from 1897–1901 and were re-classed as composites around 1910-13, with either one or two compartments upgraded to first class. In this form they were allocated SR Diagrams 323 or 324 and all three vehicles were incorporated Into Central section suburban sets until withdrawn in 1928. The photographer also noted another five bodies (ex-LSWR and SECR) at the location on the same day, with two ex-SECR vehicles near the east beach. Almost certainly, all were prepared for grounding at nearby Lancing Works. *D Cullum*

A Billinton 30ft brake third this time, seen at Littlehampton in August 1937, expertly mounted on a raised timber platform. It has eluded identity, but was an example of SR Diagrams 176 or 177, depending on whether short or long buffers were provided at the non-brake end. The Southern number would have been in the range 3705–3749 and withdrawal would have taken place anytime between 1923 and 1929 – again having be employed at one end of a Central section suburban set. Eight are recorded in the registers as being sold, but this one is still clearly on or adjacent to railway property. Denis Cullum failed to note it, so presumably the body failed to survive the war. *E R Lacey*

At Pagham on 6 September 1949 we see a bungalow that goes under the name of *Sunshine Express* – a somewhat whimsical name given the origins! At first glance, it appears to have been built from a couple of four or six-wheeled coach bodies, but appearances can be deceptive. The presence of glazed toplight windows indicates that the coaches were once 48ft or 50ft bogie block vehicles that were built for London area suburban services by Birmingham RCW Co in 1900–01 and formed into twenty seven-coach sets. These originally had frosted coloured glass in the toplights, with the class designation thereupon – blue for first class, red for second and off-white (that perhaps faded to yellow) for thirds – allowing passengers to quickly identify their appropriate compartment. Most were rebuilt into electric stock from 1928 onwards and only a very few were actually scrapped or sold off at that time. Exact identification may only be guessed at, but the section facing the photographer consists of four third-class and one-and-a-half first-class compartments (on the left), making this a Diagram 332, 333 or 334 former second/third composite. The remainder of the coach (two-and-a-half first-class compartments) has been turned into the pebbledash-rendered portion seen at 90 degrees on the right of the picture. Those few that remained as steam-hauled stock were withdrawn in 1930–34 and presumably this was their date of grounding. A couple of others were nearby. The second-class compartments were reclassified first after 1913. *D Cullum*

Billinton bogie stock dates from 1894 onwards – generally 48ft long until 1905, when 54ft became the norm until the last were constructed in 1924. This is an example of the most numerous LBSCR bogie coach – a 48ft eight-compartment third to SR Diagram 64 – of which 154 were built between 1894 and 1905. This was SR No. 2083 (formerly LBSCR No. 1211), built by Brown, Marshalls in 1895 and running until February 1933; latterly in 'long' ten-coach set 867, based at Eardley Road sidings (Streatham) and used for excursion and special traffic duties. According to Denis Cullum's notes, the body was situated on the edge of marshes at Gillingham, Kent, and could be reached by going through some allotments adjacent to the railway. As may be seen, it now looks semi-derelict in this September 1948 photograph and no doubt demolition followed soon after. The Isle of Wight steam railway recovered a similar body from Runcton (near Chichester) in 1991, SR No. 2065, and has now restored the vehicle on a 50ft ex-SR scenery van underframe. An extra 2ft of bodywork was added to allow wheelchair/disabled access to the end compartment. It is now in service carrying the fictitious SR number 2403 (the next number, 2404, was allocated to a Brighton 54ft third on the Island). *J H Aston*

Many readers will be familiar with 48ft 6-compartment first SR No. 7598, preserved on the Bluebell Railway and recovered from West Chiltington in 1989. This is similar to coach SR 7581 grounded at Bishopstone Beach after 1930 – one of a number grounded here. This July 1939 picture encompasses many of the alterations made to grounded bodies – namely a rendered brick plinth within which the body stands, then tongued and grooved board cladding up to waistline and a timber/felted roof over and covering both this body plus another mounted about 10ft behind to give living space in the area between. Note that the door droplights remain usable. Coach 7581 began life as LBSCR first No. 82 and was built by Lancaster Carriage & Wagon Company in 1894. It was allocated SR Diagram 514 and ran until September 1930. *E R Lacey*

At Bishopstone Beach again in July 1939 we see ex-Royal train clerestory saloon SR No. 7972, built at Brighton in 1897 as LBSCR No. 564 – one of the five-coach Royal train set. So, on this occasion if the owner claimed that Queen Victoria had travelled in the carriage, it may well have been true! Another Royal saloon (SR No. 7970) was also grounded nearby. Coach 7972 was allocated to SR Diagram 644 and was demoted off royal duties by the Southern Railway and formed into 'City Limited' set 925, then running between Eastbourne and London Bridge until withdrawn in March 1934, providing the well-heeled of the town with a most comfortable daily commute to London. Probably a bit different to nowadays! *L E Brailsford*

Construction of 54ft arc-roofed stock began in 1904 and, apart from the 'balloon' high-roofed vehicles of 1905–07, this would set the standard for the remainder of the Brighton's independent existence. This unique 'luggage third' was a rebuild dating from 1907, utilising four compartments from a six-wheeled, five-compartment third and four compartments from a similar second-class coach – with the resultant 4ft gap in the centre filled by a luggage compartment. Numbered as second 305, it became third class No. 1676 in 1912 and SR No. 2334 to Diagram 73 at Grouping. It was one of a number of unique (or almost unique) rebuilds dating from 1907–10, when many six-wheeled Billinton coach bodies were remounted on bogie underframes in an economical attempt to modernise the LBSCR's carriage stock. This was a loose strengthening coach and does not appear to have been allocated to any permanent set formation. It was withdrawn in July 1940 and grounded as seen at Plumstead. The New Cross Gate foreman painter's list tells us that it was painted grey – which certainly looks correct. The photograph dates from August 1952. *D Cullum*

Many 54ft brake thirds were rebuilt using six-wheeled bodies mounted on new underframes. This is SR Diagram 199 five-compartment brake third coach No. 4000, grounded in Dartford goods yard in June 1940 and seen there on 9 December 1950. It was formed in 1908 using all five compartments of 30ft third LBSCR No. 1243 (whose number the new coach retained), plus most of 30ft guard's van No. 10. This was quite a common configuration and, according to SR Diagram 199, there were eighty-four such vehicles. However, it is known that at least five did not conform to the diagram in appearance – and there might be more! This was a not uncommon feature of ex-LBSCR 54ft rebuilds. See my recent book *Southern Coaches Survey,* Section 9 (Crecy Publishing, 2019) for further details. Coach 4000 ran at one end of four-lavatory SR set 874 on Central section services until 1931; it was then reduced to a three-lavatory set on South Western section local services until around 1935, after which the set was further reduced to two coaches only until withdrawn in 1940. *D Cullum*

Opposite top: A Diagram 347 composite coach, SR No. 6218, seen in Earlswood goods yard on 21 September 1946. This was built new in 1912, as LBSCR composite |No. 125, and later ran as the centre coach of SR three-lavatory set 770, between two Diagram 197 brake thirds until withdrawn in December 1940 – by which time the set had been strengthened to eight coaches, probably for troop train work. The New Cross Gate foreman painter's list says it was prepared for grounding at Earlswood in March 1942 and, by the look of it, was used for storage. Note that the windows have been partly obscured/blocked out – perhaps for blackout restrictions and/or blast precautions. The compartment configuration, from left to right, was: first, two lavatories (back to back), four firsts, third, lavatory with short side corridor, third, so only four compartments had access to 'facilities'. *D Cullum*

Bottom: The LBSCR built large numbers of 54ft nine-compartment thirds for use as strengthening vehicles between 1907 and 1921. There were actually two variations – one built new in 1910–12 and again in 1921 and the other a gas-lit rebuild of 1907–10, utilising a six-wheeled third and four new compartments on a new 54ft underframe. Both types differed only fractionally in dimensions and you would need to look closely to spot the differences. All were taken for conversion into electric stock trailer units in 1925–26, most never receiving their allocated SR steam-hauled carriage stock numbers. Electric trailer coach 9088 is seen grounded at Fratton in the late 1950s, having been used originally as a Home Guard hut from September 1943 onwards. Built by Metropolitan as LBSCR No. 1360 in 1911, it was scheduled to become SR No. 2250 in 1923 but was instead transferred to electric stock in November 1925, and allocated to two-coach trailer unit 1094. As a steam-hauled coach it was allocated SR Diagram 71, changed to 723 as an electric vehicle. Interestingly, several later Isle of Wight conversions look identical, but were the result of a further round of rebuilding during the mid-1930s. One such conversion survives today on the Isle of Wight steam railway. *Dr T Gough*

Down to Earth Part 3A

We regret that again space has defeated us concluding the grounded bodies of the LBSCR. Part 3B, the second instalment, will appear in SW54.

51

One Evening in May

George Reeve

Merchant Navy Pacific No. 35023 *Holland-Afrika Line* backs slowly out to make the short journey to Nine Elms about a mile beyond Vauxhall Station. No. 35023 had arrived earlier with a train from Weymouth in July 1966. *John Eyers, courtesy South Western Circle*

Most railway enthusiasts have a story to tell about their time by the line side. Excitement, camaraderie, call it what you will, there was something special about spending time with friends with a common interest. For us, that is Les Hewitson, Les Tibble, Lucian Kmiotek (his father was a Polish immigrant) and myself, our journeys to 'train spot' in Woking was an adventure we embarked upon most weeks after school during the summer months. Why Woking? I hear you say. Well, for us 'Townies' it was a short steam-hauled trip to the country to watch main-line trains come and go at a major Southern junction. All manner of services came and went; electric multiple units to Guildford, Alton, Portsmouth etc., while steam trains both stopping and express, as well as freight, seemed to be passing every few minutes to Bournemouth, Weymouth, Salisbury and the far west.

One Evening in May

Woking in August 1966. The Merchant Navy *Port Line* has paused briefly with the 12.14 Waterloo to Bournemouth Central train on platform 4. Just beyond is Woking signal box, an iconic reminder of the Southern's flirtation with art deco architecture. *John Eyers, courtesy South Western Circle*

One evening in May 1965 the four of us set off to Streatham Hill station to catch the train to Clapham Junction and on to Waterloo. The last lesson of the day on Tuesdays was laughingly called 'Optional'; basically we all sat around talking about music or doing homework. However, the benefit was that on most occasions we were able to leave early, hopefully undetected. Leaving early was not the only danger we encountered – we regularly risked the wrath of the guard who inspected our cheap day return tickets, warning us that if we travelled again outside the permitted time we would be for the chop – a kind, colloquial phrase that essentially meant ejecting us at the nearest station – yes; and we believed he would! We still managed to survive many journeys without capture. Steam was disappearing fast over the country but, of course, on the Southern we still had two years of bliss enjoying main-line steam until 1967.

As we arrived on platform 2 in a four-car suburban EMU set (4-EPB) a train left in a flurry of steam and smoke. At the head was West Country Pacific No. 34027 *Taw Valley*, which was slipping like mad even though the empty stock loco was pushing as hard as it could at the rear. The train was for Salisbury, another of our destinations later that summer. Our trips were always steam-hauled and Les T, our timetable man, would always select a first stop Woking train, fabulous, and all crowded into the first coach behind the tender taking it in turns to hang out the window. There was always time to stand at the end of platform 11 and watch the toing and froing of empty carriage stock and the arrival of locos from Nine Elms shed, the great Southern edifice, only a mile or so away from Waterloo.

53

The countless toing and froing of trains from Waterloo was mesmerising with suburban and main-line EMUs and steam and diesel arrivals and departures. In addition there was a continuous stream of ECS workings to Clapham Junction in the hands here of standard 2-6-4T No. 80133, also acting this day in July 1966 as station pilot. *John Eyers, courtesy South Western Circle*

By 1965 the M7 0-4-4Ts had long gone from the ECS work, being replaced by the well-proportioned standard 2-6-2Ts and on duty this evening were Nos. 82006, 82026 and one of the larger standard tanks, No. 80154. Merchant Navy Pacifics No. 35006 *Peninsular & Oriental S.N.Co.* and No.35030 *Elder Dempster Lines* had recently arrived with trains from Salisbury and Weymouth. Les H managed to convince the driver of No. 35030 that he was studying for his GCE in metalwork and would be grateful if he could explained the finer points of driving a steam loco – 'can all my mates come on too' was met with a modicum of reluctance but nevertheless we were all able to crowd on and listen to the marvels of top link driving.

Steam West of Exeter had ended in 1964 with the introduction of the new timetable on 6 September but us chums had been lucky enough to witness the last official run of the *Atlantic Coast Express* on 4 September 1964 – a Friday I recall that meant missing a games period at Priest Hill playing fields in Ewell – it must have looked suspicious as we all turned up later at school with similar excuses at about the same time! The reason for mentioning this epic event was that some Exeter and Salisbury services were starting to be diesel-hauled. From 1 August 1964, D800 series 'Warship' diesels started taking over from steam on Waterloo to Exeter workings. The ending of the Waterloo to Plymouth timetable on Monday, 7 September 1964, saw passenger services on the ex-SR line going no further than Exeter St David's.

There was of course an exception – the 1.10am newspaper train to Plymouth, which had a through coach from Waterloo attached. The Brighton to Plymouth train and local services remained throughout from Exeter, but on 6 May 1968 the section of line west of Okehampton was closed to passenger traffic although the line to Ilfracombe lingered on.

Platforms 9–12 were popular haunts for train spotters at Waterloo and a couple are wandering aimlessly around pencil in hand. Two more senior lags gaze at No. 35026 *Lamport & Holt Line*, of 70G Weymouth Shed, as she prepares to leave with the Bournemouth Belle on 17 September 1966. To the right is Warship D818 *Glory*, by this time almost definitely on an Exeter train, while our friend from 25 May 1965, Standard 2-6-4T No. 80154, waits patiently for the next job. *John Eyers, courtesy South Western Circle*

Two likely lads wander up along an empty platform 11 in August 1966 to see what goodies await. *John Eyers, courtesy South Western Circle*

A clean West Country No. 34100 *Appledore* on the 12.35 Waterloo to Weymouth service on 7 September 1966. *John Eyers, courtesy South Western Circle*

Woking viewed towards London from platform 5, the bay platform 6 to the right. This was essentially the spot we were relegated to when a more officious member of staff was on duty. The through line is to the immediate right of the signal box. *John Eyers, courtesy South Western Circle*

One Evening in May

Right: **A glorious sight as S15 No. 30834, on a goods train in October 1961, blows off on platform 2. There were two yards at Woking used almost twenty-four hours a day. On the down side is the general freight yard, while on the up side the more interesting engineers' siding, which would see a variety of engine types including, and noted on several of our visits, Q and Q1 0-6-0s.**
John Eyers, courtesy South Western Circle

Below: **A familiar friend from May 1965. Class 5 4-6-0 No. 73117 on a Waterloo to Bournemouth train in August 1966 on platform 4. Parcels and general goods were dealt with on platform 5 and 6; any goods offloaded on the other platforms being transferred across via the lifts visible by the towers on the footbridge.**
John Eyers, courtesy South Western Circle

One of the great pleasures of Woking was observing the occasional express thundering through the station at speed. Here Battle of Britain Pacific No. 34056 *Croydon* heads the 9.25 am Weymouth to Waterloo in August 1966. *John Eyers, courtesy South Western Circle*

Merchant Navy Pacific No. 35027 *Port Line*, another of our 25 May friends, drifts into Woking on an August afternoon with the 12.14 pm train from Waterloo to Bournemouth Central. *John Eyers, courtesy South Western Circle*

One Evening in May

The summer 1965 timetable saw that the Saturdays-only through services from London to Ilfracombe were diverted from Waterloo to Paddington. We didn't realise then that by the following year that diesels would be in total control of the former LSWR lines to Exeter. Two of the D800 series locos were sitting in the sidings next to Waterloo A signal box. These 'Warship' diesels from the dreaded Western Region were throbbing away gently belching a hideous black cloud of diesel fumes across the platforms. Nevertheless, we warmed to these machines with their liveries of green and on occasion maroon, although I noted this evening that both were in Green, D819 *Goliath* and D822 *Hercules* – *Goliath* we were to see again later at Woking.

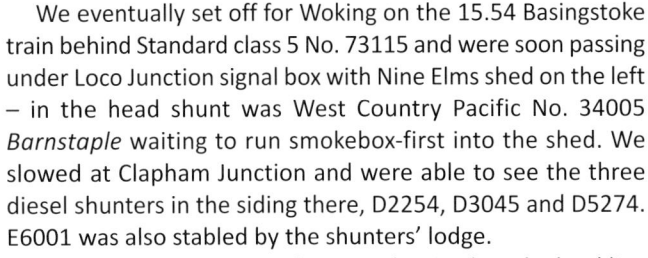

We eventually set off for Woking on the 15.54 Basingstoke train behind Standard class 5 No. 73115 and were soon passing under Loco Junction signal box with Nine Elms shed on the left – in the head shunt was West Country Pacific No. 34005 *Barnstaple* waiting to run smokebox-first into the shed. We slowed at Clapham Junction and were able to see the three diesel shunters in the siding there, D2254, D3045 and D5274. E6001 was also stabled by the shunters' lodge.

We made Woking in good time and arrived on the local line platform 5 at 16.17, immediately making our way to get some tea and sweets – I can't remember if this was on the station or indeed just outside but whatever the refreshment was most welcome. Woking was a huge station with six platforms and a Glasshouse signal box, which exists still to this day. There were two major marshalling yards at Woking, the downside for general freight and an up yard for the engineers' and permanent way. Both were busy this evening, with S15 No. 30834 working the engineers' side while U Class No. 31719 shunted the down sidings. The yards were worked round the clock, the engineers' sidings being particularly busy at weekends, of course. The record of what went on this evening is by no means complete as for some of the time we were either walking to a vantage point over the main line at York Road, the famous Tin or Twin Bridges that spanned the line at this point, or in the nearby Wimpy!

Left: **The up local platform 2 with an EMU for Waterloo.**
John Eyers, courtesy South Western Circle

Below: **Two views of Waterloo to Salisbury trains leaving Woking in May 1961. Top: No. 35016** *Elders Fyffes* **showing behind the extent of the engineers' siding and, below, another Merchant Navy, No. 35006** *Peninsular & Oriental S.N. Co.* **with the Guildford line just visible behind. The views are taken from the Twin Bridges, which again gave a different dimension for us train spotters.** *John Eyers, courtesy South Western Circle*

Depending upon which station staff were on duty, we were either allowed to stand on platforms 3 and 4 and experience the full power of an express roaring through at speed, or indeed were ushered to a safer watching spot on platform 5 and 6. The great advantage of Woking was that one could see for miles in either direction; to the north a long straight section and to the south a gentle curve. It was possible to calculate the speed of an approaching steam train by simply observing the blasts of smoke as it passed under the two bridges to the north. About a mile away was Sheerwater Road Bridge, while Monument Bridge was about half that distance. As the train passed under the bridge smoke would billow out as it blasted the underside of the arch. Counting the seconds between blasts, one could tell what speed it was doing – a simple game but it kept us occupied!

Les T kept a timetable of the comings and goings that evening, which is reproduced below. The duty numbers would, of course, have told us everything we wanted to know but alas in those days we didn't have the necessary information. We were able to log where the trains had come from by asking the drivers and firemen, but that proved a little difficult when the train didn't stop! Referral to a trusty timetable filled in the gaps … I hope. The notes from my old train-spotting note books are over fifty years old, so your indulgence would be appreciated when it comes to exact times and locations.

The Log

Time	Destination	Locomotive	Direction	Platform
16.57	Ex-Salisbury	D831	Up	3
17.03	Weymouth	35027	Down	4
17.30	Salisbury	34087	Down	4
17.42	Ex-Bournemouth	35021	Up	3
17.50	Weymouth	73117	Up	Log incomplete
18.00	Bournemouth	73043	Down	–
18.02	Bournemouth	34012	Up	–
18.13	Salisbury	D819	Down	–
18.19	Ex-Bournemouth	34109	Up	–
18.30	Exeter Central	34056	Down	–
18.40	Basingstoke	73093	Down	–
18.45	Ex-Salisbury	D822	Up	–
19.00	Weymouth	34102	Down	–
19.14	Ex-Bournemouth	35019	Up	–
19.25	Salisbury	34100	Down	–
19.35	Basingstoke parcels	73082	Down	–
19.41	Ex-Bournemouth	35020	Up	–
19.59	Weymouth	35023	Down	–
20.02	Ex-Weymouth	34017	Up	–
20.20	ECS	73081	Up	–
20.28	Ex-Basingstoke	76018	Up	–
20.37	Basingstoke	73092	Down	–

We left for home on, of all things, an EMU from Portsmouth on what appears to be the 21.02 arriving at Waterloo at 21.46. There was just enough time to see a diesel in Wimbledon yard, D6585, and D6533 and D6546 at Clapham Junction. The final part of the log was to note 76065, 76018, 80143, 82011 and 82033 at Waterloo. A marvellous day repeated many times over that summer. Oh yes, and Lucian hadn't done his homework again!

Tuesday, 25 May 1965
Waterloo–Woking

Loco No.	Class	Name
30834	S15 4-6-0	
31791	U 2-6-0	
34005	WC 4-6-2	Barnstable
34012	WC 4-6-2	Launceston
34017	WC 4-6-2	Ilfracombe
34027	WC 4-6-2	Taw Valley
34056	BB 4-6-2	Croydon
34087	BB 4-6-2	145 Squadron
34100	WC 4-6-2	Appledore
34102	WC 4-6-2	Lapford
34109	BB 4-6-2	Sir Trafford Leigh-Mallory
35006	MN 4-6-2	Peninsular and Oriental S.N. Co.
35016	MN 4-6-2	Elders Fyffes
35019	MN 4-6-2	French Line CGT
35020	MN 4-6-2	Bibby Line
35021	MN 4-6-2	New Zealand Line
35023	MN 4-6-2	Holland-Afrika Line
35027	MN 4-6-2	Port Line
35030	MN 4-6-2	Elder Dempster Line
73043	5 4-6-0	
73065	5 4-6-0	
73082	5 4-6-0	
73081	5 4-6-0	
73092	5 4-6-0	
73093	5 4-6-0	
73115	5 4-6-0	
73117	5 4-6-0	
76018	4 2-6-0	
76065	4 2-6-0	
80143	4 2-6-4T	
80154	4 2-6-4T	
82006	3 2-6-2T	
82011	3 2-6-2T	
82026	3 2-6-2T	
82033	3 2-6-2T	
D819	B-B 4 Warship	Goliath
D831	B-B 4 Warship	Monarch
D822	B-B 4 Warship	Hercules
D2254	0-6-0 Shunter	
D3045	0-6-0 Shunter	
D3274	0-6-0 Shunter	
D6533	Bo-Bo 3	
D6546	Bo-Bo 3	
D6585	Bo-Bo 3	
E6001	Bo-Bo Electro-diesel	

A final view of the approaches to Woking and the glorious Tin/Twin Bridges by York Road. No. 34102 *Lapford* approaches for a stop with the 11.10 am Bournemouth to Waterloo train in July 1966. Note also the stone hoppers from Meldon Quarry in the background.
John Eyers, courtesy South Western Circle

A Gosport Accident and the Eastleigh Footbridge

I suspect that, like many, I have an almost morbid fascination with accidents. Please do not think it is with any wish to see gore or anything like it, but more to consider the aftermath and consider cause and effect. Probably this is also why I find the Railways Archive website so fascinating. If you do not subscribe (and I would recommend doing so – it is free), even a cursory glance and I think it unlikely you will not find something of interest. Visit www.railwaysarchive.co.uk

So, plug over, in their June 2020 missive there was reference to an accident, more like a fortunate escape, near Gosport that made for interesting reading. In the same file was another, perhaps more interesting, piece on Bishopstoke (later Eastleigh) and this goes well with the well-known early image of Eastleigh reproduced here. Both incidents occurred in 1860 exactly one calendar month apart and were investigated by Col. W Yolland. https://en.wikipedia.org/wiki/William_Yolland

We start with that on the Gosport line:

I have inquired into the circumstances which attended an accident that occurred, on the 25 January 1860, to a child of 7 or 8 years of age at a level crossing rather more than a mile distant from Gosport Station on the London and South Western Railway. The level crossing in question is an occupation crossing only, the road leading from two farms to the village of Brockhurst.

There are gates across the road on each side of the line, and a stile for foot passengers to get over, at one side of the gates. The occupiers of the two farms and the gateman at a public level crossing in the employ of the London and South Western Railway Company have keys to these gates which were not locked on the afternoon when this accident occurred. The public level crossing is somewhat less than a quarter of a mile distant from this occupation crossing – too far however for the gateman to have any control over the small amount of traffic which is said to pass along the occupation road. A platelayer was standing about 70 yards from this occupation road, when the 3p.m. train from Southampton to Gosport approached the crossing at its appointed time and when it had got to within about 50 yards of the crossing and was travelling about its usual speed, perhaps 20 miles an hour, he saw a little boy on the road trundling a hoop. The boy was inside the gate when he first saw the little fellow, and he thinks the child got over the stile and did not pass through the gate. He noticed the child strike the hoop and then saw it run across the line after the hoop, just as the train reached the crossing. The engine struck the child which fell on the road, clear of the line it had crossed, and was taken up insensible and remained in that state for more than a week, but I understand it is now hoped that he will recover. The most extraordinary circumstance is that the child was not killed on the spot.

The child and its mother were on their way, from one of the cottages at the farm at which they lived, to the village, and the little boy, having got his hoop, ran on in advance before his mother, while she was getting ready to leave her house. It does not appear that there is any blame attachable to any of the Company's servants, and I know of no way by which accidents of this kind can be avoided. *W Yolland*

Bishopstoke station looking north. No date but we can now be certain it is post-1860 with the pedestrian footbridge, the bridge nearest the camera present. We also know the four-track layout was present in at least 1860 and possibly from the opening of the branches in either direction to Gosport and Salisbury. One of the eight regular shunting horses is also visible.

Next to Bishopstoke (Eastleigh):

I have inquired into the circumstances which attended an accident to a lady at the Bishopstoke station, on the London and South Western Railway, that occurred on the 25 February 1860. The Salisbury and Portsmouth Branches of the London and South Western Railway, join the main line at Bishopstoke station, where a large establishment of porters, pointsmen, shunters, and gatemen are kept, for the purpose of conducting the large traffic which arrives at and departs from this station in the course of each day, amounting to about 40 passenger and 20 goods trains arriving, and the same number departing.

The station is provided with up and down platforms on opposite sides of the line, and there are four lines of railway between these platforms, which must be crossed on the level by all passengers arriving from the before-said branches or the Dorchester line who intend to proceed on by the other branches, and in a slight degree, by the main line. Besides the trains which I have enumerated, a pilot engine and eight horses are kept at this station to do the shunting required, which will much more than double the regular train traffic passing and repassing between the platforms. On the day in question about 3p.m. the porters etc. were engaged in forming up into one train intended for the Portsmouth Branch, the two trains which had arrived from Southampton and Salisbury. Both trains stood about 50 yards apart, on the down line next to the down platform, the Salisbury train stood a few yards on the Southampton side of the crossing, between the platforms, and the Southampton train further on from the crossing. A porter had just uncoupled, or was in the act of uncoupling, the last second-class carriage of this train, when the engine in front of the Southampton train pushed back that train, in obedience to a signal given to the driver by the pointsman on duty, and struck the Salisbury train rather hard, causing the second-class carriage that had been uncoupled to be put in motion; and as this was done, a lady and her daughter who had arrived by the Salisbury train and were going to Winchester, stepped down off the down platform, for the purpose of crossing to the up platform, in order to get into a carriage for Winchester, and were both immediately knocked down by the second-class carriage. The mother fell between the rails and the carriage passed over without in any way injuring her.

The daughter fell across the rails and the wheels on one side passed over her body and seriously bruised and injured her, breaking some steel bands that formed a portion of her dress, and these inflicted severe wounds. She is now I understand gradually recovering. The next carriage to the uncoupled second-class carriage stopped about one-and-a-half yards from the place where these ladies had been knocked down. The whole of the rest of the passengers who had got out of these two trains, amounting to upwards of 20 persons, had crossed to the opposite platform before this accident took place. It does not appear that there is any blame attachable to any of the Company's servants, and there is no reason to question the sufficiency of the establishment kept at this station.

The accident must be regarded as one of a class sure to occur, at some time or other, sooner or later, when there is anything faulty in the arrangements. No accident from persons crossing the line on the level had previously occurred at this station since it was first opened, a great many years since. The London and South Western Railway Company have thus received an unmistakeable notice, fortunately not attended with fatal results, that the crossing of passengers from one platform to the other, on the level, should no longer be permitted, and I have much satisfaction in stating that I understand the directors have recognized the necessity of erecting a bridge, and the only doubt that remains is, whether it is to be an over or an under bridge; either would provide for the public safety, but I think an over-bridge will be found most suitable to the locality. (We may imagine that the poor girl was wearing a crinoline, whose support was usually made of steel and whalebone, and could be pretty substantial. The period around 1860 was about the height of the crinoline fashion.)

The Esher Slotted Signal

More from David Wigley

No sooner had the ink dried on the article in *SW51*, than David Wigley was in touch with some wonderful new information on the piece. (Page numbers refer to those in *SW51*.)

Picture page 58. This is of the down side of Esher station looking north along Station Road. Virtually everything visible has now disappeared except bridge No. 55, whose spans across the road were renewed in 1955. The single-storey building to the left was the booking office and booking hall. Beyond the two-storey structure at ground level was the bottom landing of the staircase to the down local platform 3. Esher must have been the last LSWR station to retain the numbering system where platforms were numbered, not platform faces. Above at platform level was the porter's room. Unusually, all the regular accesses to the station platforms were by staircases directly from the street, and the platform numbers can be made out above the pavement. The large enamel sign on the platform parapet is long gone; similarly the running-in board above. The timber platform awnings and platform buildings were gradually reduced and demolished in the 1960s and 1970s. Removal of the sign exposed a builder's plate, 'Joseph Westwood 1888'. This must relate to the earlier steel bridges installed in 1888 when the line was widened to four tracks just at Esher station, which were replaced in 1955. Of this structure, only the side spans supporting the platforms remain. (Joseph Westwood of Millwall, East London, were structural steel contactors favoured by the LSWR.)

The text on page 59 and following refers to the hidden slotted signal and the handling of race traffic to Sandown Park, which is conveniently adjacent to the station. As stated, the bulk of the race traffic was from London. The arrivals were dealt with easily, as racegoers leave the train and immediately quit the station. On the down local platform 3 there were two dedicated ramps open only on race days leading to a path on the South side of the railway accessing the race course. One set of gates is visible in the picture on page 63. If a train used the down main platform 2 there was an additional exit down steps to the same path visible between LSWR lattice platform fencing also on page 63. (Exmouth Junction concrete paling fencing in later views). The steps led to a subway, 55A, which passed under all tracks.

Bridge No. 55 at Esher. The crest reads 'Joseph Westwood & Co. Limited. Engineers & Contractors, London, 1868'. *David Wigley*

Handling the traffic towards London at the end of a race meeting was more difficult as everyone leaves the course at the same time. They need to access a platform, assemble and board a train, hence the special arrangements described in the article. South Western Railway still has this problem coping with return traffic from Twickenham rugby ground. At Esher the departing traffic problem was handled by the provision of dedicated race platforms 1, 2 & 3. The writer regrets that he has no information as to how the race traffic was handled before electrification. After 1937 it would appear that only race platform 1 facing the up local line was used supplementing the main station. As described, this involved introducing the hidden slotted signal Esher West 47. At the same time, Esher East signal box would have been switched in. The mechanical slotting restored to normal placing the control of signals 37, 38, 36 & 45 under East Box and making its up local to up main crossover available. This made the temporary signal West 47 the section signal for the up local line between West Box and East Box. Then trains stored in close order on the up local line at signal 51 could be admitted to the race platform to clear the crowds. The exception to this was that timetabled services would only use the regular station platform 1. After 1936 signals 37, 38, 34 and 45 had the customary SR approach light added to give distant indications for light signals at Hampton Court Junction. These only illuminated when the semaphore arm above was cleared.

With regard to race platforms 2 and 3, the use of these in the absence of any information of working before electrification remains a mystery. These platforms were not electrified. As the photographs show, there was only one signal for two tracks in the accepted layout for two stop arms on the same post and both clearly marked 3. The post is situated in advance of the convergence of the tracks for race platforms 2 and 3. The lower arm 'TO THRO' refers to the direct connection to the up through line, which can be made out in the illustrations on pages 62 and 64. This was a nasty layout, coming as it did in the middle of a line speed super-elevated curve. Later pictures show it to be the first part of the installation to be abolished.

On non-race days, Signal 51 was Esher West's up local home signal and signal 50 applied to the crossover from the up main, No. 25. As far as I know there was no facing connection from the up local line to the race platforms. There was a rarely used trailing connection to the up siding controlled by a ground frame, 'West Points'. At some there must have been a direct connection from the up siding to the up through, as it appears on the OS plans of 1913 and 1938.

To expand on the handling of returning crowds, the subway 55A adjacent to the West Signal Box gave access to all platforms either by ramps or staircases. Additionally, there was another entrance with further staircases to race platforms 1, 2 and 3 at their country end. A gate and a race traffic booking office in Lower Green Road plus an arch 55B under the race sidings led to these.

In the writer's experience, race platform 2 and the extended up siding were only regularly used to hold two loco-hauled ten-coach special traffic trains. Race platform 3 saw hardly any use, although I can remember an occasion when some horseboxes were unloaded here, not race horses but destined for the Metropolitan Police mounted training establishment, which still continues at Imber Court. The ramp to the road still exists.

What remains today? I regret I did not keep a chronology of all the changes at Esher station over the years. Esher East signal box was abolished in September 1961, which would have ended the requirement for the special signalling arrangements. The site of the box can still be seen on the up side adjacent to bridge No. 54. Signals 35, 37, 45 and 36, plus the facing crossover up local to up main, were placed under the sole control of the erstwhile West Box. This made race platform no. 1 within station limits. The goods yard on the down side retained three sidings plus two short spurs until it

Interior Esher signal box. Sykes 'lock and block' and illuminated diagram are visible. *David Wigley*

Close-up of the 'lock and block' instruments and the all-important (but potentially lethal if used incorrectly) release key.
David Wigley

closed in December 1962. The race platforms survived until the end of steam traction on the main line and have since been demolished and the staircases backfilled. The remaining signal box was abolished in March 1970, its functions greatly reduced and replaced by the then new Surbiton panel box. The station buildings have been replaced by a small flat roofed structure on the former loading bank adjacent to the down slow line next to the goods shed, which survives. A footbridge has been provided to access all platforms but the fast line faces are now unlit and abandoned. The subway 55A is now available at all times and forms the step free access to the station. The track layout is just four plain lines, all switch and crossing work long forgotten. There is still traffic on race days but this is handled by adding special stops for the Alton and Basingstoke services.

During my experience all race traffic was handled by electric multiple units. There was a single exception. The *Daily Mirror* newspaper on one occasion chartered a special train to bring celebrities to the race course. The local enthusiast population got excited, thinking they were going to see the LMS compound painted yellow. However, all that arrived was one of the local standard Class 5s. It would be an event if that happened today.

The slotted signal that was the subject of the original article had a sad end. It was accessioned by the Curator of Historical Relics but during dismantling it was smashed and destroyed.

...and Speaking of Signals...Three Curios...

Bridge No. 55 at Esher. The crest reads 'Joseph Westwood & Co. Limited. Engineers & Contractors, London, 1868'. *David Wigley*

BRITISH TRANSPORT
COMMISSION
THE USE OF THE
COMMISSIONS
PREMISES AS A
THOROUGHFARE
IS PROHIBITED

Above and opposite: **Underslung starting signal at Ashford, Middlesex.** The position of the station canopy created the need for this installation that we have to admit, while necessary, was also unduly complicated.

Right: **Finally Fareham West and the starting signals for the line to Cosham (left) and Gosport (right).** Here the conventional approach has been taken with co-acting arms to afford maximum visibility. Note the arms applicable to the Cosham route were also slotted from Fareham East box.

Remember *Remembrance*, April 1922 to June 1935

Gerry Nichols has kindly submitted this selection of images of No. 333 during its time as a tank engine and prior to rebuilding. Images are from the archives of the Stephenson Locomotive Society.

LBSCR No. 333 was one of five engines ordered from Brighton soon after the 1918 Armistice but was only completed as the last of the batch in April 1922, the delay caused by a backlog of repairs. The engine is seen here in photographic grey and with the name given as a memorial to fallen LBSCR men.

Remember *Remembrance*, April 1922 to June 1935

In November 1932 '2000' was added to the number by the Southern Railway as part of the scheme to distinguish engines within the three SR constituents. (Former SECR locos had '1000' added to their numbers, while LSWR stock retained their original 1, 2 or 3 digit identification. This was in place of the prefix that had been used, respectively 'A', 'B' or 'E', and representing Ashford, Brighton and Eastleigh.)

In smart SR livery, the addition of the name precluded the addition of any ownership details on the tank side and instead this was shown above the number on the bunker.

71

No. 2333 dwarfing its train: the oval buffers allowed for better contact with stock when pushing on a sharp curve and so reduced buffer locking.

In charge of the prestige 'Southern Belle', notice the Pullman cars in the early livery with the cantrail painted white. After electrification of the Brighton line in 1933 and the consequent replacement of the locomotive-hauled 'Southern Belle' with the electric 'Brighton Belle', all of the class of 4-6-4T were transferred to Eastbourne, where they were the mainstay of the express link until 1935. An earlier attempt to utilise the class on the Mid-Sussex route to Bognor fell foul of the chief civil engineer, who refused to allow them regular passage over the Arun bridge at Ford. With electrification also due to reach Eastbourne, the decision was taken to rebuild the class as tender engines with the designation 'N15X'. The name *Remembrance* was also retained. In revised form, and now as BR No. 32333, the engine continued in service until April 1956, when the influx of Bulleid Pacifics and Standard types rendered the class redundant. Sadly No. 333 was scrapped; she would have made an ideal candidate for preservation. (Further detail on the class will be found within *Locomotives of the LB&SCR* by Bradley, published by the RCTS.)

The Southern on Social Media
Pigeons and People
Roger Simmonds

The term 'social media' may be one that is commonplace in the twenty-first century but it was certainly not something heard on our 'steam and electric' railway years ago. Like it or loathe it, this form of communication is here to stay and while we perhaps look in astonishment at the younger generation as they communicate more it seems via smartphone than any other means, I have to admit (speaking as a partial Luddite) it can have its plus points as well.

As an example of this, my long-term friend and research colleague Roger Simmonds alerted me to two recent postings both relative to Winchester but both also exactly the sort of thing he knows we like to feature in *SW*. The first relates to the release of pigeons and the second to the railwaymen of yesteryear, hence all I can say is, 'Roger, it is over to you …'

(Quality and consequently reproduction size may not be all we might always wish, but as a record both are unique and dare we say we also welcome other similar contributions.)

All of a flap at Winchester

Pigeon racing first became popular in Belgium during the nineteenth century, when the modern racing pigeon, the Homer, started to be bred for its speed and ability to fly long distances. As the popularity of the sport grew, other nations became interested in racing pigeons, including the United States and Great Britain. The first formal pigeon race in the UK was held in 1881. The National Pigeon Association become the governing body.

Above: **'All in a flap' in the down side loading dock at Winchester.**
Courtesy Greta Chapman

Right: **Perhaps a case of 'take cover'. The birds would soar high and through the wonders of nature relate their location, a case of 'bird nav', before making off to their respective lofts often hundreds of miles distant.**
Courtesy Greta Chapman

Unlike today with advanced timing equipment available, pigeon racing in previous years was quite different as it was more difficult to determine race distances and times. The pigeons would be released by railway guards or porters at railway stations and when they reached home, their race rings removed and rushed to the nearest post office for the race time to be registered officially by the postmaster. Soon after this, people started to use their own racing clocks and remove the pigeon's race leg ring at the end of races to record their own times.

It formed significant business for the railway companies and although the LNER was the biggest carrier for this type of traffic, closely followed by the LMS, the Southern had its fair share. This was made possible by the passing of the Transit of Poultry Order of 1919. The traffic could be as large as whole trains run as homing pigeon specials or as one or two vans attached to passenger or parcels trains. Parcels and utility vans were converted in some cases with the provision of shelving to enable the stacking of the wicker baskets. Smaller transits could just utilise the floor of parcels vans. The LNER had dedicated vehicles for transporting pigeons, some being conversions of ex-NER vehicles. These had drop-down shelves for ease of access. Some pigeon specials run could be formed by as many as sixteen vehicles. The first of these ran on the NER in 1905. To give an idea of volume, there were 23,982 birds sent by rail to the 1907 Up-North Combine race.

On the SR general utility vans tended to be the most common form for moving the birds to their predefined release point. Smaller sets of releases could be handled in the guard's compartment of passenger trains, perhaps limited to two or three baskets. Release points would generally be from a station passenger platform or a loading dock if present. The important consideration was that there were no overhead obstructions, such as electric, lighting or telegraph wires, and that there was no activity nearby to distract the birds as they were released.

After the birds had flown there was the tidying up to do. *Courtesy Greta Chapman*

The railway staff involved *had to* carefully note the release time and date on the label, then return the empty baskets on the next available train. There were strict instructions about looking after the birds and also about ensuring precise release times where large numbers of baskets/birds were released as they were not supposed to be all set off close together. Staff were also given strict *instructions* not to feed them because they would not fly home,

As far as the Southern was concerned, regularly used stations designated for pigeon release were Weymouth, Templecombe, Bournemouth and indeed Winchester, where these images were taken. In 1929 the LMS alone claimed it was carrying seven million birds during the racing season and running as many as seventeen special pigeon trains on some days, including those heading south taking the Somerset & Dorset line for release at Templecombe and Bournemouth,

Admiring or perhaps worried looks...? *Courtesy Greta Chapman*

In 1929 the *LMS Railway Magazine* noted that: 'At the time of its inception, some 40 years ago, pigeon racing was considered to be essentially a working man's hobby, but this is no longer the case, and many of the newspapers in their account of the Bournemouth [north-west combine] race, stated that the owners of the birds taking part in the flight comprised members of all ranks of society.'

So maybe not quite the stereotyped 'Wally Batty' cloth-cap image that comes to mind.

The Southern on Social Media

and others via Basingstoke for release at Winchester. It was claimed in 1900 that probably half a million people were interested in the result of a race of Yorkshire clubs released from Winchester having arrived there by special train.

The outbreak of the Second World War curtailed pigeon racing but, of course, many of the birds had a vital role in the conflict. Over a quarter of a million pigeons were used in the last war alone. British pigeon fanciers donated thousands of their best birds to help in the war effort and all three of the armed services plus the civil defence used pigeons for carrying messages.

The rail transport of pigeons for racing resumed after the war and continued to be a useful source of income for British Railways, although increasingly road transport became more popular with breeders and fanciers as the motorway network increased. Pigeon Specials had all but disappeared by 1966. British Rail, despite still earning the annual revenue of around £150,000 in 1975, took the decision to end the freight carriage of pigeons as well as of most other livestock; a proposal that came into force on 1 July 1976.

Greta Chapman also sent some images of her father Les Grist, again at Winchester taken between eighty and ninety years ago. Les worked his entire railway career at Winchester, starting with the LSWR in 1917 aged 16 as a junior porter rising to full porter status during the 1920s with the SR. He became a porter-signalman by the 1930s and signalman grade shortly after. He had a spell away during the war, joining the RAF, but returned at Winchester by the end of hostilities. As well as the Southern station during the 1950s, he worked shifts at Winchester Chesil box and on rare occasions at Micheldever.

For reasons that are unclear, during the latter 1950s he ceased to be a signalman and took up a post in the parcels office before finally becoming a ticket collector. He was well known at the station by regular passengers for wearing a rose in the jacket buttonhole in this later role. He retired early in 1959 and passed away in 1964 aged just 63.

This time it is the release from just a few baskets. Porter Chapman is in charge in the Winchester station goods yard. *Courtesy Greta Chapman*

Above: **Les Grist at Winchester c.1925. He is standing on the up side – the station signal box may just be made out at the far end of the opposite platform.**

Left: **Moving forward a few years, this is Les c.1936.**

Above: **Alongside the signal box. The role of porter/signalman meant he was qualified to work the box and might well do so as a relief between shifts or perhaps to cover short notice absenteeism.**

Above left: **Might it have been his handiwork with the chalk on the board…?**

Left: **Les greeting VIPs, whose identity was not recorded.**

Below: **Winchester station staff, c.1936. Names would be welcome … Do you have some images of staff tucked away? We would love to see them and perhaps similarly feature them in** *SW*.

Colour Interlude

Roger Holmes

It is a long time since we have included some colour imagery from Roger Holmes – far too long in fact.

No excuses, just a simple 'I must get around to that ...'! Well no more excuses this time and as this has been a primarily steam-orientated issue we felt it only right that the selection from Roger should be orientated towards other traction. All were taken in the period 1965–67 (Roger did not often record actual dates but we do at least have the month, year and location.) South Western division, of course, but no shame in portraying diesels, electro diesels and REP/TC sets, especially at a time when most were instead turning their cameras to focus on what remained of steam. Indeed the period 1966–67 through to the early to mid-1970s has a dearth of colour material. Perhaps then a few memories will be stirred from a man who knew a good photograph, and a good location, when he saw them. Best of all, they are also all on Kodak film.

Early foray of a Class 33 and certainly what appears to be a new TC unit at Southampton Central in July 1966. At this stage the original canopy on the Platform 1 was still intact – it was replaced soon after. The lack of passengers could well indicate a test trip.

The '16' headcode indicates an Alton to Southampton Terminus service seen here having just left Northam in August 1966 and with the now long-lifted west chord that allowed direct access from the Terminus to the Central station on the left. The lines on the right ran between Northam yard and Southampton Terminus goods, and also afforded access to the Chapel tramway.

D6531 having just left Lyndhurst Road for Bournemouth, also in August 1966. Here the conductor rail and pots have been installed, while on either side of the running lines, sidings have been lifted.

Colour Interlude

Displaying headcode '75' but hardly a Portsmouth and Southsea to Salisbury via Southampton service, as at that time the line between Portsmouth and Southampton had not been electrified while that between Southampton and Salisbury never has been. But we do have a 4-EPB followed by what appears to be two 2-EPB sets (coaches five and six in blue livery with 'double arrow' symbols) just west of the rebuilt Redbridge viaduct and heading in the direction of Bournemouth in December 1966. The sets were also probably on a test run.

The early months of 1967 witnesses another Class 33 near Woodfidley in the New Forest with what could well be a further test run with at least one TC set in tow. The 'C2' headcode – also seen earlier – is not confirmed.

79

This time the '85' headcode (also Portsmouth to Salisbury via Southampton) could well be correct as it is a TC set being propelled by a Class 73 just south of St Denys in February 1967. (The signal box of the same name may be seen just under the footbridge.) Many years earlier there had been a level crossing here but this was removed at a very early date. On the extreme left are the headshunts and loco stabling point for Bevois Valley yard. TC set No. 422 was later strengthened to four vehicles.

Clean diesel, dirty steam engine. Both outside the front of Eastleigh shed in March 1967. (The 'Bulleid' is not identified.)

Steam and diesel again, this time in opposite directions near Bishops Dyke close to Beaulieu Road station and again in the New Forest in April 1967. The Standard Class 4 is not identified, while the Class 73, we think E6002, is in propelling mode.

Another eight-car EMU formation, this time with a pair of two-car EPBs leading, the first No. 5761, attached to a four-car set. The location is Sway bank just west of Lymington Junction in April 1967, with the train heading for Bournemouth. EPB units were (fortunately) not seen regularly on stopping passenger turns on the Bournemouth line but on the few occasions they were there were invariably complaints about the lack of toilet facilities by passengers who had by then become used to VEP sets.

We conclude with a TC set emerging into the sunshine at the west end of Southampton Tunnel in May 1967. This time it is TC No. 407 with the description for a Waterloo to Western Docks service. The orange curtains were a feature of the REP and TC sets for the Bournemouth electrification and contrasted well with the original bland all-over blue that was first applied.

81

Rebuilt
The Letters and Comments Pages

Another goodly crop of most interesting and appreciated views from readers this time. In no particular order but starting with **David Wigley** (whose additional notes on Esher have appeared earlier in this issue). David writes: 'More mundane matters, in *SW51* p.36 the surmise is correct; BGP – Bogie Gangwayed brake van Pigeon. These vans were fitted with fold-down shelves to receive pigeon baskets. P.78 top. I do not believe this train is going anywhere near Guildford. It is on the down Brighton through and the headcode is Victoria to Littlehampton via Hove. The formation is interesting. It appears to be three 4-RES units in temporary form. The kitchen diners have been removed and replaced by trailers from PUL or PAN units, the centre unit receiving a Pullman car. P.82 bottom. The Newhaven boat train is on the down Brighton through, about to pass Haywards Heath.'

David also adds some extra information on items within *SW49* and *50*. SW49 cover: '… down Portsmouth & Southsea stopper near the end of the climb to Buriton tunnel. Engineer's staff platform for up line just visible on extreme left. Probably Easter holiday as the headcode 57 has run independent of the Alton service and no leaves on trees. *SW50* p.7: This must be during the 1955 footplate strike; Clapham junction middle of the day, no engine in sight and the yard crammed with stabled vehicles. Included are at least four of the eleven six-coach Bournemouth sets, which would normally be out on the road. *SW50* p.38: Higham, towards London, fits the headcode displayed, the leading unit is a 2-EPB.

Reverting back to John Click again in *SW52*, readers of the series will recall mention of the tests related to the fire-throwing propensities of the original Bulleid breed. No. 34033 was mentioned but at the time we had failed to locate an image. Of course, we had just gone to print when this one arrived. Agreed SR and not BR days and at the time nameless, but still rather pleasingly brand new at Brighton. *Jeremy Staines*

'P.86 Will the real "Battersea" please stand up? The cabin has a GW appearance and the track in the foreground is GW. It must therefore be the SB on the WLER, which existed just west of the overbridge for Battersea High Street at Battersea station. It was a simple block post between Latchmere Junction and Chelsea Basin SBs. Appears on 1916 OS plan and was abolished in March 1936. Finally, p.89 right: Regrettably this picture is nothing to do with Frant, It is instead of Faversham LCDR looking towards London before the rebuilding of 1897.'

At this point we will also bring in **Colin Duff** again on the subject of **Esher.**

'A belated congratulations on reaching edition 50, and now 51.' *Thank you, I will admit others have expressed similar comments but I can only echo the words I have said previously, which are – it is also only with your help that any of this is possible – Ed.)*

'In *SW51* on p.58, I think the picture is of the down side of Esher Station, and not the up side as in the caption. Very familiar territory for me. My uncle (a former Nine Elms driver) bought the Station House (behind the cameraman's left shoulder) off BR and renovated it. My Aunty Anne and Cousin Jill still live in it. The current station's forecourt is where the shoe repair shed is in the photo and the abutment wall is still there, as is a mirroring wall on the left which was part of the station building under the canopy to Station Road. A further identifier is the four-arm signpost that can be seen the other side of the bridge in the distance, this marking the junction of Station Road (in the opposite direction at that time back to the A3, now the A307), Weston Green Road (towards Thames Ditton), Ember Lane (towards East Molesey) and Lower Green Road (roundabout towards Esher). You would see no cross road if you were looking under the bridge from the up side.

'The Station House has been prone to trespassers walking through its garden and climbing over the garden fence to get into Sandown Park Racecourse for nothing!'

From **Chris Longley.** 'Only just got hold of *SW50* and a bit of a feast on the S & D from Paul Hocquard. However, I think the image on p.50 is Binegar and not Midford. Possible clues: Midford had 17 levers, Binegar 24 (21 visible in image). The door is at the wrong end of the operating floor to be Midford and also the gable ended roof is in place rather than Midford's flat roof. The Cat at Bath (Green Park) I presume to be a 'Midland Red'! P.52: I do so much remember parcels and mail by passenger train and I believe this image was taken at Shepton Mallet, the service would most likely be the 3.40pm Bournemouth– Bristol, which called at Shepton at 6.11pm. This train connected at Mangotsfield with the overnight train to Newcastle and so every effort was made to run to time.'

Next from **Simon James**. 'As always *SW51* was an interesting issue across many parts of our interest. I am sure I am, however, not alone in noting that in the picture on p7.8, 4-RES 3067, is one of those re-formed as a 4-COR(N) unit with a 6-PUL trailer third replacing the restaurant car, while the second unit of the train is one of the short-lived 4-PUL units. The headcode 16 is to Guildford via Epsom, which may indicate a Sunday diversion as there is no headcode to Pompey via this route that I am aware of. On p.81 I think that in the back platform at Herne Bay is the same train seen leaving on p.80, the roof of vehicle 2 in the train leaving the back platform seems too similar. The Hornby's headcode on p.82 is very odd, a Vic–Newhaven train should show two discs above the offside buffer but without knowing the location it is not really obvious what this train might be. On p.83 perhaps E5015 has performed a shunt release for the E5xxx, which can be seen at the buffer stops of the other low-level platform.

'Very interesting article about the singling of the WoE Main Line, I'm sure we would all like to thank the ***** person in Railtrack who managed to replace the former double track bridge at Templecombe with a single line bridge as well as the truly inspired (!) way they have redeveloped the station currently when all that was needed to improve matters for passengers was to extend the double track through the already existing if disused down platform and move the points to the other end of the station, the usual lack of foresight. Looking forward to the next issue already!'

Can anyone please add some notes on the re-forming of the EMUs mentioned into 4-COR(N) and 4-PUL sets please – Ed?

Now from **Alastair Wilson. '**Quite a change – and a great pleasure – for me to read the latest edition (referring to *SW51*) of *SW* without knowing what was coming (*Alastair is one of our abused proof readers!*) – mind you, I'm not suggesting I don't want to go on, so long as you want me to copy-edit and proof-read for you! *(Yes please!)*

'But first of all, would you please pass my thanks to Peter Swift for supplying the answer to my earlier query about D1s/E1s south of Tonbridge.

'Secondly; that image of the 'N' blowing off on the freight in the frontispiece – you say "West of Exeter" and I would agree: it looks to me as if it might be between Ilfracombe and Barnstaple – possibly on the Ilfracombe side of the summit between the two places.

'And the Pictorial from Graham Smith, and the caption on p.76 about 'Five and Nine, the Brighton Line'. I assume you know, though quite possibly many other readers don't, that it originated because 5/9- was the price for an excursion to Brighton about a century ago.'

From **Mark Brinton** re *SW51* **– various and interesting.**

'A couple of minor matters from Southern Way 51: SR 15 ton Brake Vans on p.22. There are two of these vans, Nos 55710 and 55724, preserved at Havenstreet. Both lost their sandboxes prior to transfer to the Island in 1967. No. 55724 has been 'restored' and sees regular use in the railway's historic goods trains. 55710 is stored.

Similarly the first of two additional images to accompany John Perkin's article on the Waterloo & City electrics. Here we see No. 74S on the incline to the power station at Durnsford Road...

And then an unusual view of some coal wagons on the elevated section.

'With regard to the incident on p.27, it is unlikely that the brakes would leak off on all four carriages sufficiently to cause it to run away. A more likely cause is the EP Brake miniature circuit breaker tripping, or the equivalent fuse blowing. This would then cause all the brakes to release. The EP Brake is not a fail-safe system and without the relevant train wires energized it will default to release. The unit appears to have escaped with relatively light damage, considering the demolition job it has done to the buffer stops!

'The LNER Pigeon Van on p.36 is coded "BGP" on the end of the vehicle. Ian's understanding of the code is correct as in the Southern Region Appendix to Carriage Working Notices for May 1970 it lists "BGP" as "Pigeon Van (with gangways)".

'The grounded LBSCR Brake body on p.43 (lower), was still in position a couple of days ago when I walked past it on one of my regular walks around the village. These days it looks a bit tidier than in your photo, as it has now been entirely clad in plywood. It is still in regular use by its owners for their beach gear.'

Which we really do have to follow with more from **Nicholas Owen,** who, of course, was the original correspondent on the 'runaway' and other items in *SW51*.

'May I add a few observations – and begin with an addition. It's to my article about the Caterham runaway, the early-morning 4-EPB that set off driverless – and thankfully passenger-less – from Caterham to travel unmanned until diverted into, and derailing at, buffer stops at Norwood Junction 10 miles away. I have learned since the piece was published that the driver scheduled to work the train, one Jim Longley, was delayed on his way to work. Another driver sought to help Jim have a quick-as-possible getaway by releasing the handbrake, which took rather a lot of laborious turning. The air in the cylinders leaked off, and off the unit went.

'A marvellous PS is that some seven years ago, when Jim died at the age of 84, his family arranged to be played at his funeral, yes, *The Runaway Train*!

'Elsewhere was the fascinating pictures of the Crystal Palace High Level branch, doomed to under-use once the Crystal Palace main building burnt down in 1936. (Incidentally, a blaze my late father remembered watching from his then family home in Anerley close by.) The platform layout inside the terminus train shed was interesting, with two of the roads having platforms on either side to aid the flow of passengers to the Palace complex back in pre-fire days. I didn't know the siding layout at the High Level was so extensive, including a couple of electrified roads.

'Another incidental fact. It was on The Parade beside the High Level station that I recall in about 1954 or 1955 seeing my first Routemaster, turning across the wide road at the end of its cross-London run on route 2.

'Finally, the picture of a 2-EPB at Sanderstead, p.81. It can't apparently make up its mind whether it's coming or going. My guess is the guard had been quick to pop the tail lamp on, and the driver either hadn't got round to changing the "3" headcode for the shuttle run to Elmers End, or didn't want to be bothered with doing so every time the short trips ended. The train is in the Down platform at the then-limit of the third rail. The crew would swap ends, and with the guard riding "shotgun", the driver would reverse back past a country-end crossover before regaining the up platform. The "10 car" marker is explained by the operation in earlier days of commuter trains to and from Charing Cross (headcode 28) and Cannon Street (29), which would also be shunted as above.

'Interesting that both lines are seen electrified. In the days before the extension of the third rail to East Grinstead, the electric shuttles only used the down before retracing their steps over the crossover beyond the bridge. In earlier days, there had been through commuter trains from Charing Cross (headcode 28) and Cannon Street (29), so I suppose it's possible they were shunted into the up platform before departure. Anyone know?'

We have also been continuing the theme of the **architecture on the Chessington branch** article in *SW51* with correspondence between **Alan Postlethwaite** and **Stephen Spark**.

Stephen asks first, 'Was the Chisarc type of canopy used for any other station in Britain?' Alan responds, 'I know of just one – at Manchester Oxford Road, built post-war. I don't have any photos of it but I do have one of a Chisarc canopy in Rotterdam which I photographed in 1961.'

Stephen also continues, 'That is very interesting; I believe there was one at the Ocean Liner Terminal in Southampton too, but I haven't seen any photos that provide much proof of that. Naturally most pictures concentrate on the loco and coaches rather than constructional details of the platform (photographers are funny like that!).

'The shell construction style was fairly short-lived, as the same effect – spanning a large space without intermediate supports – could be done far more cheaply, though less stylishly, with steelwork once that material was "off ration" for the building trade. And shell construction required skilled labour, which was in short supply post-war, plus it took longer and therefore probably cost more.

'If you haven't done so already, do take a look at Stockwell bus garage – a good example of what the technique could do and reminiscent of some Continental railway stations of that period. If the war hadn't broken out when it did, I wonder if Ellson might have opted for something like that for wherever the next major station rebuilding was to have been (and where, incidentally, would it have been – what were the next major projects the SR had in mind in 1938/39?). And what, I keep wondering, would the Southern have done when the Chessington line reached Leatherhead? Might it have become a showpiece like Surbiton but with fully integrated Chessy-style canopies? One for the "fantasy" modellers, perhaps! Somebody has indeed made a model "south of Chessington South" station called Chalky Lane. Personally, though, I'd have used that as an excuse to include Robert Goddard's planned NG line that was to have started at Chessington South and paralleled the SR line as far as Chessington Wood then crossed the road (on a bridge?) before terminating at the zoo. Just imagine that: futuristic stations, Southern Electric trains, narrow-gauge steam and tigers!'

Next from **Ray Grace** on Deal and *SW51*.

'Thank you for publishing my letter on the Betteshanger Colliery operation at Deal. On p.33 of *SW51* you mention that you have been unable to find an image of Deal station; if you have a copy of the Middleton Press *Dover to Ramsgate* South Coast Railways book you will find in images 45 and 46 two good photographs taken from the station footbridge. Image 45, looking in the "up" direction, shows the direction along which the brake van was "fly shunted" past the "up" starting signal and returned by gravity to the centre road. By the way, my parents (and I) lived in the house with the signal projecting over the garden and my bedroom window was the dormer above the van on the bridge! The centre siding has the short goods in it and the "Birdcage" forms the stock of the Deal to Minster (Thanet) shuttle service that operated throughout the day.

Chisarc canopy photographed by Alan Postlethwaite at Rotterdam in 1961 *– just 300+ miles and approximately six and a bit hours from Chessington...*

This page: At the start of this issue we included a photograph of a G16 4-8-0T shunting at Feltham. Thanks to our friend Amyas Crump we also now have a copy of one of the little pamphlets produced by the Southern Region on the location. (The SR did similar for the various works when there was an open day, does that mean there was ever such a thing at Feltham Yard? The scale of the complex can be gauged from the fact from entrance to exit was only just short of one mile. (For a full history of Feltham Yard we can recommend *Monograph No. 8 Feltham Concentration Yard*, published by the South Western Circle; www.lswr.org/services/monographs.html

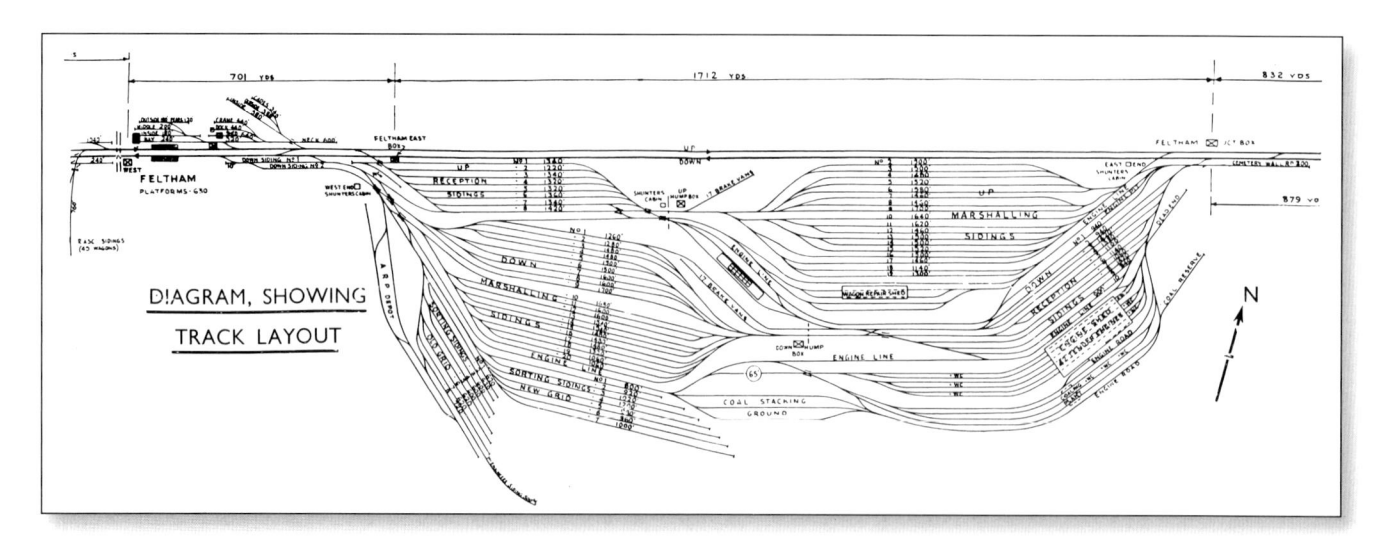

'If we now look at the lower image (46), in addition to the caption we can see the "C" Class, which had been shunting the yards all morning preparing to leave to return to Ramsgate MPD with a single brake van, the other wagons appearing to form a separate train! There is also the top of a "Q1" to be seen on the turntable behind the rake of coaches and the Minster shuttle loco, probably a LM Class 4 2-6-4T, in the process of running around and backing on to its "Birdcage" stock. This dates the image to pre-1958.

'There are a number of unusual features in these photographs. The rake of crimson/cream stock is neither of the resident nine-car sets (217/218) normally berthed there and it was unusual to see a line of wagons in the adjacent long siding, which was usually occupied by one of the "resident" sets.

'As a matter of interest, the SR "art deco" signal box is the prototype for the recent Hornby model. Also, the loco shed (used for wagon repairs) is not as long as it looks in the photograph. The lighter-roofed shed is across the road behind the loco shed and was the "East Kent" bus garage.'

From **Graham Buxton-Smither,** on **John Click, the Croydon Tangle** and a new topic**, 'Tipping Drivers'.**

'I have come late to the party yet again but have just finished reading *SW50* and it proved another that was read cover to cover in one sitting (well, lying down actually). *(But I do appreciate the thought – Ed!)*

'Although I enjoyed the entire issue, two items grabbed me in particular: the superb piece featuring the memoirs of John Click and the article on The Croydon Tangle. First-person accounts are always interesting and often very revealing – you can almost feel as if you, the reader, were present at each instance covered in his recollections. As someone who used to travel six days a week as a schoolboy to East Croydon from Reigate and thence back to South Croydon and vice versa, the history of Croydon and its environs was always going to be a winner. The one thing I wished I could find was the photo that I know exists of No. 916 *Whitgift* at South Croydon being clambered upon by some over-enthusiastic Whitgiftians; the school is only a short walk from the station.

'I was reminiscing with an old chum last week and we both remembered our parents/guardians doing something that no longer happens – tipping the train driver. I recall arriving from Southampton in the late 1950s and my uncle giving me a half-crown, telling me to walk along the platform and hand it to the engine driver while thanking him for a smooth and timely journey. I vividly recall a very well dressed, elderly lady handing some coins to the driver as I approached; he was standing by the cab as if expecting this. I also remember my uncle's gaze as I started back after a very polite 'Thank you young sir' from the driver – it only occurred to me later that he was probably checking to see that I hadn't simply pocketed the coin … My chum was a regular passenger on the Western Region and remembered his father doing the same thing two or three times a year at both Paddington and Swansea after a good journey. I had quite forgotten the tipping until we had the conversation and meandered down Memory Lane.

'It was a gentler and more respectful world back then – at least in some aspects of common courtesy.

'Oh well, enough reminiscences! I'm waiting on Issue 51 – deeply frustrated that the scheduled publishing date of 1 July has come and gone but that always serves to heighten expectations …'

(The subject of tipping the driver is something we would be glad to hear more on from other readers – Ed.)

Comments also received and appreciated from Tony Francis and David Morgan, some of which points have previously been discussed in 'Rebuilt'. Other correspondence has been held over.

The Editor reserves the right to edit/precis letters as necessary although nothing will be done to change the meaning of the subject. We may also decline to print some letters for reasons of space or hold these over until sufficient space is available. Finally, the usual caveat. The views expressed may not be representative of the editorial team at Southern Way.

'The Southdown Venturer Railtour' and the End of Steam
The Last Rites of Steam and Lines in Sussex Part 2

Les Price
Images by the Author

The unusual sight of steam at Victoria but, of course, destined for the special working. The engine is backing down on to its train, hence the headcode discs and tail lamp. *Les Price*

With the closure of the Steyning branch in 1967, three routes providing through routes between London and Sussex to the South Coast had already been closed. But two were left, other than the Brighton Main Line, which itself split at Keymer Junction, providing an eastern arm to Lewes.

In each case these two provided an alternative route should problems occur on the main Brighton line. One was in East Sussex from Ashurst Junction to Lewes via Uckfield. The other was the remaining arm of the original Mid-Sussex line from Horsham to Arundel Junction. But prior to the two closures covered in Part 1 of this article, the first to go had been the Bluebell line, which was an early casualty, closing on 17 March 1958.

'The Southdown Venturer Railtour' and the End of Steam

Was it irony or indeed a portent of things to come, that it was the closure of the Bluebell line that subsequently led the Bluebell Railway to become the first standard-gauge preservation movement? The success of the Bluebell ultimately led to other lines up and down the country being reopened as 'heritage railways', demonstrating their value to the local community. Not only was part of the Bluebell preserved but it has ultimately grown and thrived to the point of suggestion that sometime in the distant future the line may reopen through to Ardingly and just possibly to Lewes once again.

The virtual elimination of duplicate north–south railways in Sussex had been recommended by the Beeching Report in 1963, which included the 'Wealden Line' from Tunbridge Wells to Lewes. This route could be broken down into two sections; the northerly one being from Eridge to Tunbridge Wells and the southern one from Eridge to Lewes. The closure of these two sections subsequently came to be something of a hot potato.

Realising the diminishing facilities available to give steam engines the ability to work over lines in Sussex, the 'Southern Counties Touring Society' arranged the 'Southdown Venturer' Rail Tour for Sunday, 20 February 1966. This was one of the last Steam Specials to operate across Sussex and more specifically the very last over the Uckfield to Lewes line.

This tour was different, inasmuch as it left from Victoria rather than Waterloo. So the participants arrived to see a rebuilt West Country Pacific No. 34013 *Okehampton* drawn up at the head of a rake of eight coaches. She had been rebuilt in 1957 and since September 1963 had been a Salisbury (70E)-based locomotive. But prior to that she had spent fifteen months at Brighton, so would have previously traversed some of the route she was about to navigate.

Leaving promptly at 9.38 am, *Okehampton* took us down the LBSCR main line through Clapham Junction and Selhurst before calling at East Croydon to pick up further passengers. We left there, on schedule, at 9.56, taking the Oxted Line at South Croydon. This line was instigated in March 1884 as the Croydon, Oxted and East Grinstead Railway. In the middle of the battleground for territory between the South Eastern Railway and the London, Brighton and South Coast, the two companies ceased hostilities, which had been raging for the previous twenty years, and promoted a joint venture.

By Riddlesdown we had left the Surrey stockbroker belt and dived into the half-mile long tunnel of the same name burrowing beneath the North Downs before emerging on to Riddlesdown Viaduct. A further viaduct was crossed at Woldingham and there were more short tunnels through the North Downs before and after Oxted, which we passed on time at 10.12. At Hurst Green the line split into two branches.

The one to the right to East Grinstead was formerly jointly owned as far as Crowhurst Junction, where there had formerly been a connection to the SER main line from Redhill, the latter continuing in an almost easterly straight line to Ashford then on to Folkestone and Dover. The line onward from Crowhurst Junction as far as East Grinstead became solely LBSCR-owned. Beyond East Grinstead it led on to the 'Bluebell Line' to Culver Junction. We on the 'Venturer' took the left fork, which here became solely owned by the LBSCR, and crossed the county boundary into Kent.

Passing Edenbridge, which even today still boasts two stations serving a small town with a population of around 10,000, and Hever, birthplace of Anne Boleyn, we arrived at Ashurst Junction. At Ashurst we took the Withyham Spur to Birchden Junction, which opened on 7 June 1914 and completed what was called the 'Outer Circle Line'; providing an alternative route between Brighton and London via Oxted. However, only bowevereing considered a subsidiary cross-country double line through East Sussex and not figuring in the Southern's electrification programme in the 1930s, it remained the last steam-operated line in the area.

Previous to this, in 1894, the 'Wealden Line' from Tunbridge Wells to Uckfield had been doubled. This line formed the east to south spur from Groombridge Junction to Birchden, making Eridge, the next station, an important junction with routes diverging to each of the four points of the compass. *Okehampton* continued to head south and a mile further on eased the train through Redgate Mill Junction, where the now rusting rails of the 'Cuckoo Line' were noted, before tackling the 1 in 75 gradient beyond.

Photo stop at Crowborough (and Jarvis Brook). We were here ten minutes, the none-too-pleasant weather not enough to discourage the photographers and observers. *Les Price*

The 'Wealden Line' had been the name given to the railway connecting Lewes with Tunbridge Wells, a distance of 25¼ miles. Now on this line, we came to a halt at Crowborough, almost 40 miles from Victoria. The train's passengers couldn't get out of the coaches quickly enough. The photographic stop was scheduled to be for eight minutes. As was often the case with trains such as this, we arrived a minute down and left two minutes late. It has to be said that in those days shepherding enthusiasts was no easier then than it would be today and ten minutes was the norm for most photographic stops.

Crowborough is an affluent town in the heart of the High Weald 'Area of Outstanding National Beauty'; it had a population of 20,600 in 2011. In 2018–19 passenger usage of the station stood at 358,000 journeys per year, so the railway has remained of critical necessity to the town's prosperity. In the late nineteenth century, Crowborough was promoted as a health resort based on its high elevation, with rolling hills and surrounding forest. Estate agents even called it 'Scotland in Sussex'! Probably the town's most famous inhabitant, Sir Arthur Conan Doyle, spent the last twenty-three years of his life here, living at Windlesham Manor.

From Crowborough the undulating nature of the line was emphasised and the brakes were applied as we descended the 1 in 75 bank into the ¾-mile long Crowborough Tunnel before crossing Sleeches and then Greenhurst Viaducts, each about 185 yards long. Having then passed Buxted, we approached Uckfield. The dominating geographical feature here had always been the level crossing in the centre of the town with the railway bisecting the High Street. The station was situated immediately to the west of this level crossing, whereas the signal box was situated to the east of it.

Uckfield is a growing town, its population expanding from 3,500 in 1931 to 14,500 in 2011. In 2018–19, 472,000 passengers used the station. The closure of the section of line south from here to Lewes itself ultimately became something of a conundrum, a word that could be considered an understatement. The last day of operation was announced for 6 May 1969. Meanwhile, the last day of services had already taken place on 23 February! How this came about will be explained in Part 3.

The entire train service between Uckfield and Lewes ceased from that date. A finer farce could not have been written! By then the travelling public had completely lost interest, so when the final trains left Lewes and Uckfield at 8.46 pm and 8.42 pm respectively no ceremonials or organised events took place to mark the passing.

The level crossing at Uckfield, however, continued to cause traffic congestion in the town for another twenty years, for no good reason. Then, in 1991, the old station closed and was replaced by a single-platform new station, on the eastern side of the crossing, so freeing up the road.

Back on board the train, south of Uckfield, 'The Southdown Venturer' crossed and re-crossed the meandering River Uck before coming to Isfield, now home to the 'Lavender Line'. One might have imagined that this sobriquet arose from memories of lavender being carried away from the station by train. The answer is much more prosaic: A E Lavender and Sons were local coal merchants who operated out of the small goods yard at Isfield. The preserved coal office is still in situ. Isfield Station and a mile-long section of the line northwards were bought privately in June 1983.

Certain covenants were put into the deeds of sale forbidding building on the line of route. A condition of the sale was that any property forming part of the formation could be compulsorily repurchased if the land was required in order to reinstate the through route sometime in the future.

From Isfield we journeyed on, dropping gently down through Barcombe Mills and then to the former Culver Junction, where the Bluebell Railway had formerly re-joined. Barcombe Mills was once popular with anglers, who came in large numbers during bank holidays to fish in the nearby River Ouse.

The small independent Lewes and Uckfield Railway had opened on 11 October 1858. The LBSCR supported this 7½-mile-long line running via the small village of Hamsey on the west bank of the River Ouse. It joined the Keymer Junction to Lewes line at Uckfield Junction 1½ miles from Lewes, north of Lewes Tunnel. The drawback was it did not allow through running to Brighton.

In 1864 the LBSCR purchased the smaller company. In order to create a through route between Tunbridge Wells and Brighton, obviating the requirement for trains to reverse direction at Lewes, authorisation was obtained to build a new line. This opened on 1 October 1868 and the original route was abandoned.

As we passed, the abandoned course of the original route could be seen heading south-west on the 'up' side. We were now on the new 1868 alignment, 3 miles long and running almost parallel with the coast line at its southern end; it enabled trains from Uckfield to gain independent access to Lewes without having to pass through Lewes Tunnel. On high embankments, first crossing a viaduct over the River Ouse, there followed a further bridge over Cliffe High Street and then a girder bridge across sidings. Descending on a 1 in 60 gradient, *Okehampton* eased around the sharp curve into Lewes, which we passed on time at 11.12 am.

Fifteen minutes had been allowed to cover the eight onward miles to Brighton, where *Okehampton* rolled into the terminus on time at 11.27. Only four minutes had been allowed here to enable a 'Crompton' to attach to the rear of the train. An emboldened note had been placed in the Itinerary, '*Passengers to remain seated on arrival at Brighton*'. The entire complement complied and we left two minutes early! With *Okehampton* now in tow at the rear, we were taken out to Preston Park.

This particular 'Crompton', a Birmingham Railway and Carriage Company BR Class 33 Bo-Bo Diesel Electric No. D6543, had been allocated, newly delivered, to Hither Green (73C) in February 1961. This was still the case at the time of this movement. The 33s were specifically built for the Southern Region. By removing a steam heating boiler and replacing it with a larger eight-cylinder engine in place of the original six cylinders of the Class 26s, it was given greater power. This complied with the Southern Region's traffic requirement at the time, which relied on tourist traffic. This was heavier in the summer when train heating was not required, the locomotive then being available for freight work in the winter.

After being transferred to Eastleigh (70D) in September 1966, it was renumbered '33025' in February 1973. Happily '33025' has survived. Variously being named *Glen Falloch* and *Sultan*, it is still active; currently based at the Steamtown Railway Centre, Carnforth, painted in West Coast Railway Maroon Livery and logo.

We arrived at Preston Park a minute and a half early for a scheduled ten-minute photographic stop. But here the tour participants reverted to type and we left a minute and a half late after all the reprobates had been shepherded back on to the train. The 'West Country' was now positioned to continue its onward journey along the coastal main line through Worthing and Chichester. After crossing the Sussex–Hampshire county border we stopped briefly at Havant to change train crew.

Fareham was the next stop, where one of our old friends, 'N' Class 2-6-0 No. 31411, which had been our power on 'The Wealdsman' the previous summer, was waiting to take the train down to Gosport tender-first. But since this article is really about lines and steam in Sussex I will be brief about the remainder of 'The Southdown Venturer'.

From the evidence of numerous trucks in the yard at Gosport, there still appeared to be a flourishing trade in domestic coal supplies and there was still a line into the admiralty yard. But this was not to last. Three years later, on 30 January 1969, the whole branch closed except for a short section at the Fareham end.

After running round the train, No. 31411 returned us up the branch to Fareham, where the Mogul once again ran round and took us down to Portsmouth Harbour. Despite her sprightly appearance, her end was nigh and she was withdrawn from Guildford (70C) just two months later. The engine was cut up at Cashmore's (Newport) five months on.

After servicing at Fratton (70F), *Okehampton* reappeared to take the train back to London Bridge via Guildford, Effingham Junction and Wallington, where a stop was made to 'set down'. While Fratton had officially closed to steam some time before, it still retained servicing facilities. The 'West Country' was not the last steam engine to visit; this fell to a Guildford-based USA 0-6-0 Tank No. 30072, affectionately known among Guildford loco men as 'Little Jim'.

On 9 July 1967, 'Little Jim' was the last engine in steam to leave Guildford, stopping at Fratton for water on its way to Salisbury for disposal. That coincided with the last day of steam on the Southern. I have confirmed this with my old mate, Mick Foster, the last passed fireman at Guildford. He was actually on duty at Guildford that day; although he was not on board. Perhaps the last days of steam at Guildford may be another story.

Like 'Little Jim', *Okehampton* survived to the end, but was withdrawn from Salisbury on the final day. Entirely coincidentally, she followed in the same tracks as No. 31411 to Cashmore's (Newport) for breaking up, in October 1967.

'N' No. 31411 at Fareham waiting to run down to Gosport. The stock used consisted of eight vehicles, some if not all of Bulleid origin. *Les Price*

No. 31411 in the process of running round at Gosport. Passenger services to Fareham had been withdrawn in 1953, although rationalisation had been going on since the First World War with the closure of the line to Stokes Bay. The latter was accessed by a triangle, part of which may still be discerned from the trackbed curving off to the left. *Les Price*

Inside what had once been the cavernous interior of Gosport station. Enemy action caused considerable damage to the structure in the Second World War and it was never rebuilt. Although listed, it took many decades before the site was anything near presentable again. The former siding to Clarence Pier diverges to the left. *Les Price*

Portsmouth Harbour and soon to depart of the final leg back to London Bridge. *Les Price*

Returning now to Sussex; the swansong of Brighton Depot (75A) had come in autumn 1963 when the final allocation of ten Bullied Pacifics, including *Okehampton*, were all transferred away. *Okehampton* went to Salisbury. The motive power provision for the continuing Plymouth–Brighton service then became supplied by Exmouth Junction (72A). The afternoon arrival at Brighton returned the following morning with the 11.30 am Plymouth, as far as Exeter. In the summer of 1964 all steam services at 75A ceased, meaning any 'failures' that needed attention at Brighton now had to be dealt with at Redhill.

During the winter of 1965–66 steam continued to work through to Brighton on the Plymouth, although by now the allocated locomotive had to lodge overnight at Fratton. This required light engine running between Brighton and Fratton on a daily basis.

Clearly the Southern Region had made a significant misjudgement when ordering the 'Cromptons', as specified earlier in this piece. There was no other diesel of the requisite power arrangement and heating facility to work this particular train and that would explain how steam received an extended lease of life on this service.

The last scheduled steam working into Brighton was the Plymouth train on Saturday, 30 April 1966; headed by West Country Pacific No. 34098 *Templecombe*. Regardless of the actual weather outside, this was the final day of steam heating requirement on the Southern Region, which, in turn, signified the end of the steam requirement.

Another last throe of steam power came with the LCGB 'Reunion Rail Tour' of 10 December 1966. Battle of Britain Pacific No. 34089 *62 Squadron*, throughout the 1950s traditionally a Stewarts Lane (73A)-based engine, took the train out of Waterloo. After the closure of Stewarts Lane to steam she was shunted out to Salisbury. It was reported that the Pacific was banked out of the terminus, with her eight-coach rake, by a BR Standard Class 5 4-6-0 No. 73020.

From Clapham Junction she took her train on a circular tour of south-western suburbs before returning there. From here BR Standard Class 4 4-6-0 No. 75075 took over and worked the section from Clapham Junction through Hurst Green and St Margaret's Junctions to East Grinstead (High Level) and then on to Three Bridges.

Meanwhile, *62 Squadron* had run down the Brighton main line to Three Bridges. From there she worked the train onward to Brighton and then back to Victoria. She continued her life working out of Salisbury before she too was withdrawn on 9 July 1967. Like many of her sisters, she followed the familiar route via Westbury to Cashmore's at Newport, where she was dismantled during May 1968.

Then the absolute final steam rail tour in Sussex came on 19 March 1967 with the SCTC 'Southern Rambler' Rail Tour, hauled by West Country Pacific No. 34108 *Wincanton*, which by this stage may have had its nameplates removed. Taking the main line from Victoria to Brighton, it covered the only other remaining lines in Sussex, except the coast line west to Chichester and the Mid Sussex, returning to London via Polegate, Eastbourne, Lewes and Keymer Junction.

No. 34108 was reported to have been in filthy condition but was a late replacement engine for the programmed locomotive, No. 34089 *602 Squadron*, which for some unexplained reason had become unavailable; presumably she had been failed on shed. Both locomotives were, by then, Salisbury based but ironically No. 34108 was withdrawn a fortnight before No. 34089; the latter again on 9 July 1967.

And that was it as far as steam working was concerned. However, the fight to keep the 'Wealden Line' open went on.

Book Review
Parsons & Prawns and *Southern Style*

Parsons & Prawns
The story of the first 180 years of the railway at Micheldever Station, Hampshire

Peter L Clarke, Dever Publications.
ISBN 978 0 9542929 4 2

Regular readers will recall we have often said it can be surprising what sometimes lies hidden among the aficionados of local history when it comes to railways. The mainstream archives may well reveal the salient facts and figures but can unintentionally bypass the minutiae from which answers to our quest for knowledge can also be gained.

This is most certainly the case in this new booklet by Peter Clarke on the history of Micheldever Station.

Put simply, the village of that name came about solely because of the railway and its one time throw-off point for traffic to continue to Andover by road, that is until the line west of Basingstoke was completed. Visit today and while the majority of sidings are long lifted and a motor car tyre company now occupies the former station goods yard, the main station building remains a time capsule complete with its all-round veranda.

Within forty pages, author Peter Clarke takes us on a concise tour of history from the earliest days through expansion and the subsequent development of the area just north of the station into the massive storage and fuel depot many will recall. Mention is also made of the facilities existing in the modern-day station. The text is enlivened by tales of associated history, and while this is a book intended to appeal to a wide audience, including how Micheldever Station saw one of the first motor cars leave here for Datchet in 1895, the railway historian will not be disappointed with several interesting images that were new to your reviewer.

Sadly Peter Clarke passed away the very day the book was received from the printer but copies are available via suebell0906@gmail.com

Produced on good-quality paper and to approximate A5 portrait size, there is a recommended donation price of £10. The facility to purchase by BACS is also available.

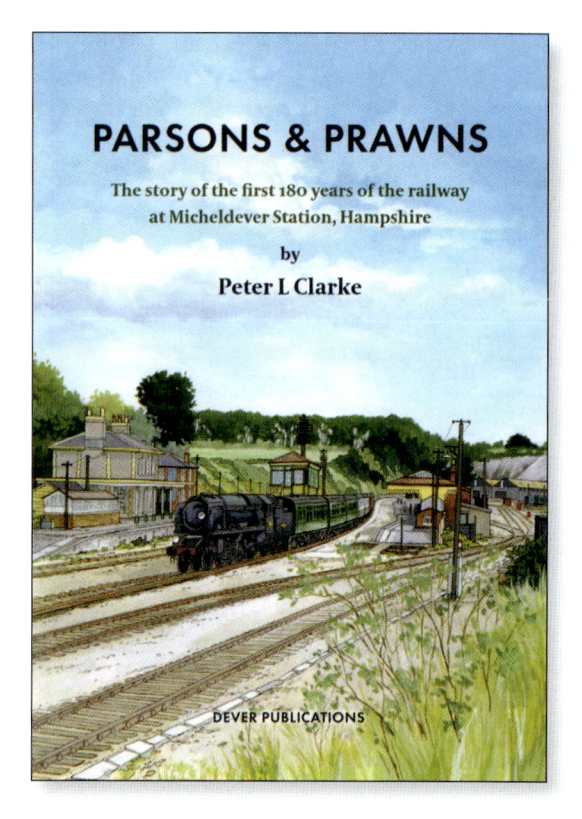

Book Review

Southern Style
The Southern Railway

John Harvey. Pub. Historical Model Railway Society.
ISBN 978 0 902835 37 5

This is the fourth volume in the series of livery books on the Southern Railway, its constituents and the Southern Region.

Like the earlier books, it is produced to A4 portrait style, printed on art paper, paperback and as with the previous LSWR and post Nationalisation works, compiled by John Harvey.

In many ways this latest volume is a joy, 240 sides and a separate folder of colour swatches make it both essential to the modeller but also a valuable source of information to the student of railway history. While at first glance it may not appear cheap at £35 (rrp), despite its worth railway books do not sell in the ten of thousands and the amount of work that has gone into compiling it will likely never be truly appreciated by those outside our hobby interest.

What makes it so worthwhile is the wealth of detail from locomotive and rolling stock liveries under both Maunsell and Bulleid and including the early changes applied to existing LBSCR and SECR and LSWR stock in the years immediately post-1923, to stations, lineside structures, rolling stock and even coach moquette colours – many of the latter illustrated in colour. Mr Harvey has left almost no stone unturned as there are also sections on the liveries of the Lynton & Barnstaple and the Somerset & Dorset, both of which, of course, had evolved their own unique colour schemes.

If I have any criticism it is in two areas; firstly I would perhaps liked to have seen a few words on the letter fonts used by the Southern both for locomotives and rolling stock but also on stations and perhaps even going so far as to cover advertising material.

Secondly, to my eye some of images are perhaps a shade dark and could even perhaps have benefited from being reproduced larger. That said, to enlarge some of the views further would also have increased the page count and with it the cost. Here commercial considerations quickly come into play but as it is probable that the market for this book will already have a shelf full of other Southern-related books, the reader will probably access some alternative illustrations from this source.

John Harvey and the HMRS are to be congratulated on this work, it deserves to succeed and be on the shelves of all readers of *Southern Way*.

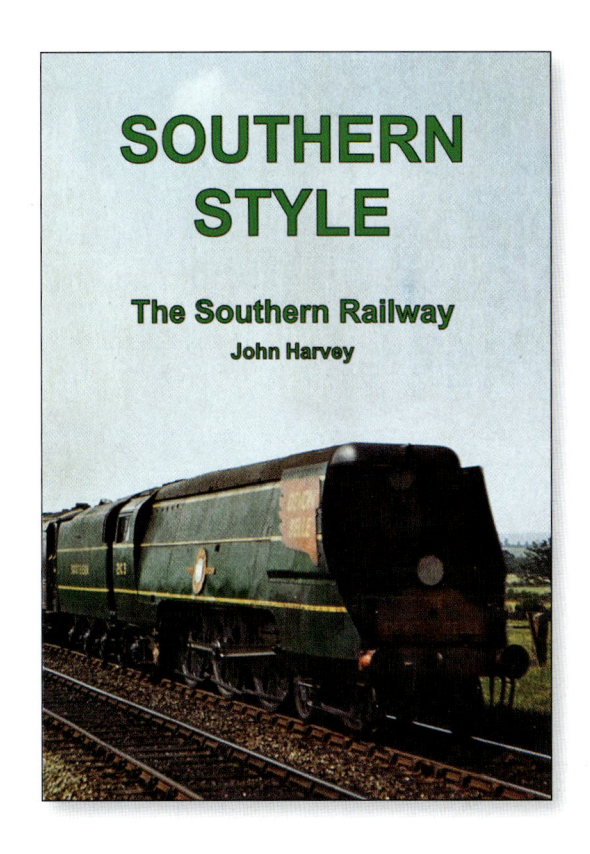

Next time, in Southern Way No. 54?

Certainly Part 2 of Alan Postlethwaite's 'Wandle Valley' and Mike King's 'Down to Earth' on the LBSCR. A further instalment of John Click, and, despite it being spring, some interesting Christmas mail and parcels workings from Richard Simmons. We are also working on two pictorial branch line histories and the intention is also to include some notes on the Southern's unique Sentinel railcar. Available in April 2021.

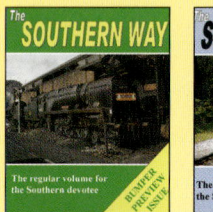

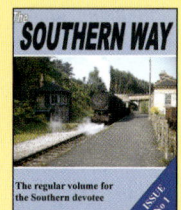

The Southern Way

The regular volume for the Southern devotee

MOST RECENT BACK ISSUES

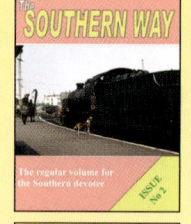

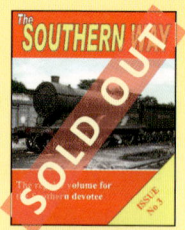

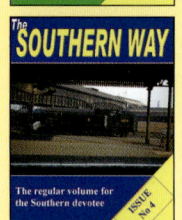

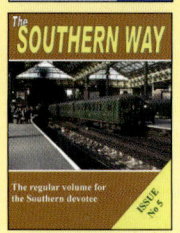

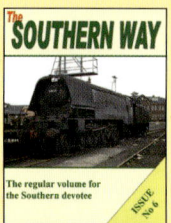

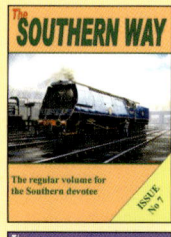

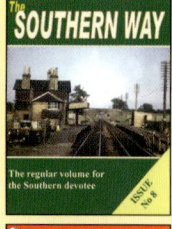

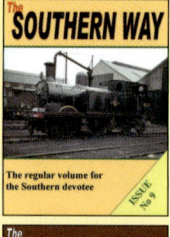

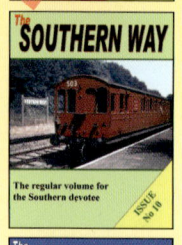

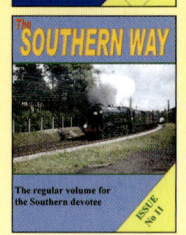

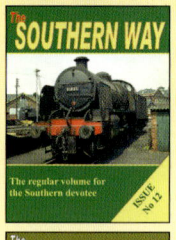

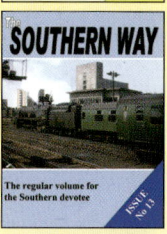

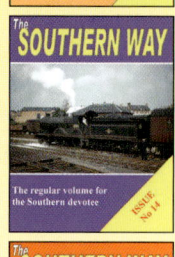

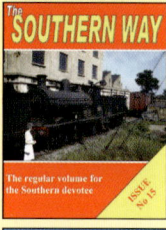

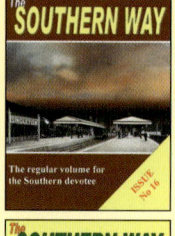

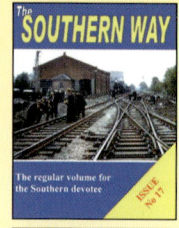

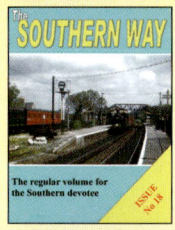

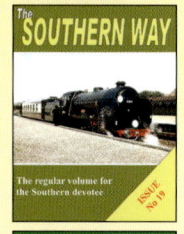

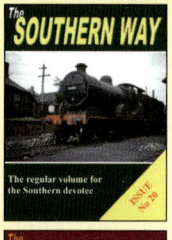

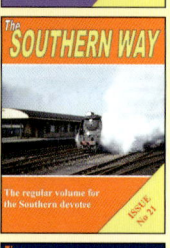

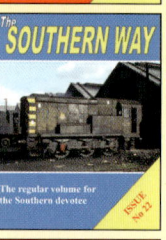

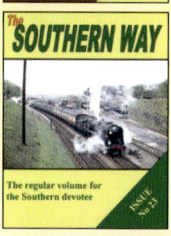

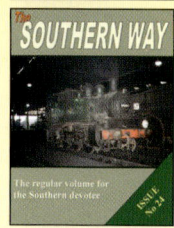

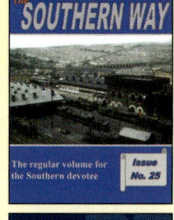

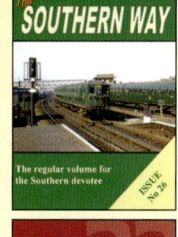

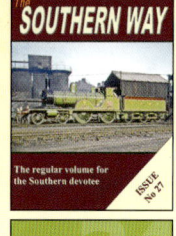

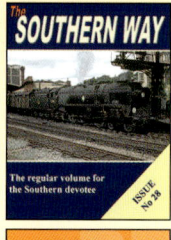

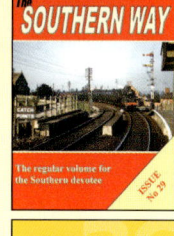

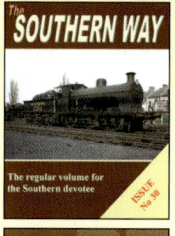

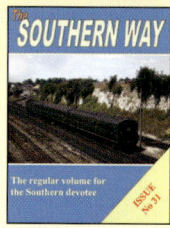

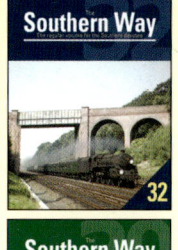

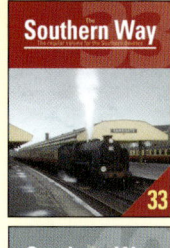

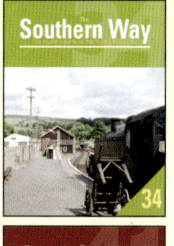

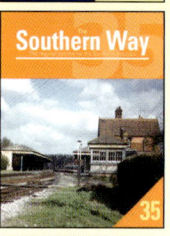

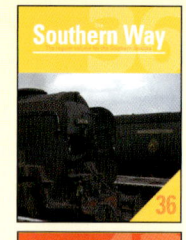

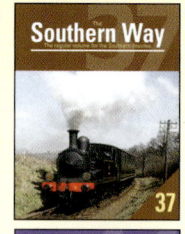

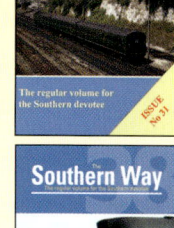

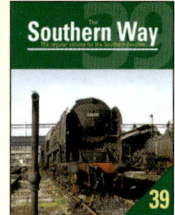

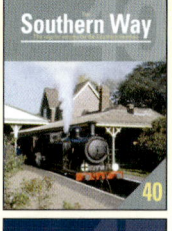

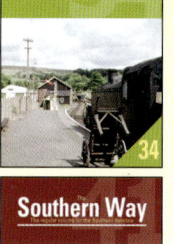

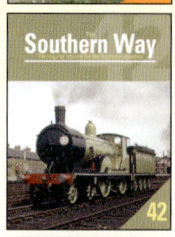

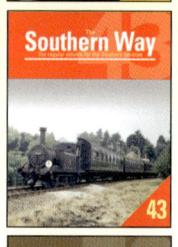

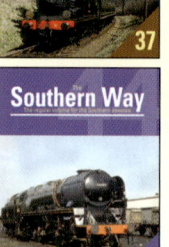

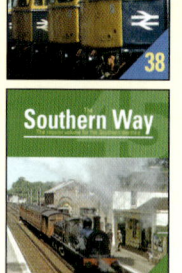

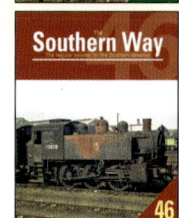

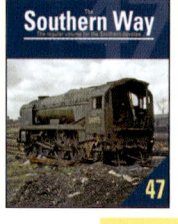

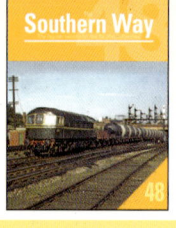

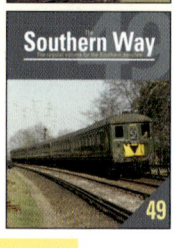

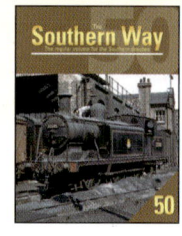

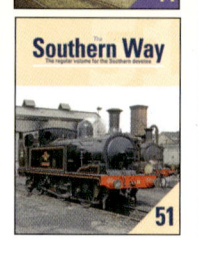

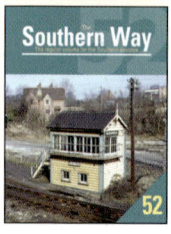

The Southern Way is available from all good book sellers, or in case of difficulty, direct from the publisher. (Post free UK) Each regular issue contains at least 96 pages including colour content.

£11.95 each
£12.95 from Issue 7
£14.50 from Issue 21
£14.95 from Issue 35

Subscription for four-issues available
(Post free in the UK)
www.crecy.co.uk